Contemporary Environmentalism in the Baltic States

This book examines the relationship between environmentalism and political transition in Estonia, Latvia and Lithuania. Environmentalism was a strong force behind the nationalist movements preceding the independence of the Baltic States and the subsequent collapse of the Soviet Union. The contributors illustrate how the Baltic States face contemporary environmental challenges through transition, enlargement and integration into the European Union.

This edited collection brings the reader through the political, economic and social changes of the post-Soviet Baltic States as a way of examining how transition influences environmentalism. Secondly, the book looks at how the enlargement of the European Union to encompass the Baltic States has influenced political and legal approaches to the environment. Finally, the contributors examine how regional and global environmental narratives have shaped contemporary environmentalism in the Baltic States. From 'eco-nationalism' to the world's first 'green' prime minister, environmentalism continues to play an important role in Estonia, Latvia and Lithuania.

This book was published as a special issue of the *Journal of Baltic Studies*.

David J. Galbreath is a Senior Lecturer of Politics and International Relations at the University of Aberdeen. He has published on several policy areas in the Baltic States including minority rights, political parties, and foreign policy in journals such as the *Journal of Baltic States*, *Cooperation and Conflict*, *Nationalities Papers*, *Geo-politics*, and *Governance*.

Contemporary Environmentalism in the Baltic States

From Phosphate Springs to 'Nordstream'

Edited by David J. Galbreath

Routledge
Taylor & Francis Group
LONDON AND NEW YORK

First published 2010 by Routledge
2 Park Square, Milton Park, Abingdon, Oxon, OX14 4RN

Simultaneously published in the USA and Canada
by Routledge
270 Madison Avenue, New York, NY 10016

Routledge is an imprint of the Taylor & Francis Group, an informa business

This book is a reproduction of the Journal of Baltic Studies, volume 40, issue 3. The Publisher requests to those authors who may be citing this book to state, also, the bibliographical details of the special issue on which the book was based

The AABS is subsidizing the price of this book by waiving its entitlement to receive royalties.

Typeset in Lapidary by Value Chain, India
Printed and bound in Great Britain by MPG Books Group, UK

British Library Cataloguing in Publication Data
A catalogue record for this book is available from the British Library

ISBN10: 0-415-57226-6
ISBN13: 978-0-415-57226-2

Contents

Notes on Contributors

Timofey Agarin is a post-doctoral research fellow at the University of Aberdeen. His work focuses on the cooperation between civil society groups and the Baltic governments on environmental as well as social integration issues.

Rune Holmgaard Andersen holds an MSc in Political Science from the University of Aarhus, Denmark. At present he is a PhD candidate at the Institute of Government and Politics at the University of Tartu, Estonia. His research focuses on political and economic transition theory, economic development and New Institutional Economics.

Daunis Auers is an Assistant Professor at the Department of Political Science, University of Latvia. He has most recently edited a book on US–Latvian relations (*Latvia and the USA: From Captive Nation to Strategic Partner*, University of Latvia Press 2008), co-authored an article on the extreme-right in Latvia and Estonia with Andres Kasekamp (*Journal of Contemporary European Studies* 2009), and co-authored, with Juri Ruus and Algis Krupavicius, a chapter on the use of referendums and initiatives in the Baltic states (Simon Hug and Karin Gilland Lutz (eds), *Financing Referendum Campaigns*, Palgrave).

Richard Caddell completed a PhD in international law at Cardiff University in 2009 and is a Lecturer in Law at Swansea University. His research work engages international law, with particular interests in international environmental law, the law of the sea and human rights, and post-Soviet legal transitions in Central and Eastern Europe, with special reference to the Baltic states.

David J. Galbreath is Senior Lecturer at the University of Aberdeen in the Department of Politics and International Relations and editor of *European Security* (Routledge). He is the author of several books on the Baltic states, including *Nation-Building and Minority Politics in Post-Socialist States* (ibidem 2005) and *Continuity and Change in the Baltic Sea Region* (with Ainius Lašas and Jeremy W. Lamoreaux, Rodopi 2007). His journal articles on the Baltic states have appeared in *Governance, Nationalities Papers, Geopolitics,* the *Journal of Contemporary European Studies*, and numerous occasions in the *Journal of Baltic Studies*. He holds a Doctorate in Politics from the University of Leeds.

Jussi S. Jauhiainen is Professor of Applied Geography and Regional Planning at the University of Oulu in Finland and Associate Professor of Urban Geography at the University of Tartu in Estonia. His research interests are urban and regional development, policies and networks, and the Baltic Sea region.

Ellen Karm received her MA in International Studies: Russian, East European and Central Asian from the University of Washington in 2006. Previously she was employed in the private sector handling international legal and business matters. Currently she is working on non-fiction and fiction writing projects.

Tove H. Malloy is Director of the European Centre for Minority Issues (ECMI) in Flensburg, Germany and Lecturer at the University of Southern Denmark. Her research focuses on the legal and political aspects of national and ethnic minority rights in international law and international relations, especially in the European context. She is the author of the monograph *National Minority Rights in Europe* (OUP 2005) as well as book chapters, journal articles and working papers.

Nora Mžavanadze is currently a PhD candidate in Environmental Sciences and Policy at the Central European University, Budapest, Hungary. She has a Master of Science degree in Environmental Sciences and Policy from Central European University and a Bachelor degree in Economics from Vilnius University. Her main research interest lies at the interactions of economy and environment combining the two educational backgrounds. Ms Nora Mžavanadze has worked on various aspects of sustainable development (assessment, strategic management, corporate social responsibility, environmental management) and her PhD dissertation focuses on environment and development issues in post-socialist countries.

Tarmo Pikner completed his PhD at the University of Oulu in 2008 on socionatural cross-border networks around the Baltic Sea. Currently he is a researcher at Tallinn University in Estonia. His topics are urbanization of nature, cultural economy and collective participations in civil society.

Allan Sikk is a Lecturer in Baltic Politics at University College London School of Slavonic and East European Studies. His main research interests are new political parties, electoral politics and quantitative methods in political science. He has previously published in *European Journal of Political Research, Party Politics* and the *Journal of Communist Studies and Transition Politics*.

INTRODUCTION: FROM PHOSPHATE SPRINGS TO 'NORDSTREAM': CONTEMPORARY ENVIRONMENTALISM IN THE BALTIC STATES

David J. Galbreath

Once the catalyst for independence movements in the late Soviet period, environmentalism in Estonia, Latvia, and Lithuania has become greatly influenced by the processes of social, political, and economic transition. With the rise of the schizophrenic response of toleration and repression to liberal reform in an authoritarian Soviet Union, activists in the three Baltic states set out to use the environment as a specific representation for what the Soviet occupation had done to the Baltic republics in general. Each of the Baltic republics had their own environmental issues with which to deal. In Estonia, environmentalists focused on the ecological damage from phosphate and oil-shale mining in the north of the country. In Latvia, the aim was to prevent another hydroelectric dam on the Daugava, a central river splitting the country in two. Finally, in Lithuania, environmentalists set out to campaign for the closure of an ageing nuclear power plant and prevent the construction of another plant. Each one of these movements to protect the environment was a move against ecological degradation, while at the same time a challenge against the central planning of the Soviet economy and overall Baltic inclusion in the Soviet Union. In the Baltic republics, the environmental movements and the independence movements were inextricably linked, to the extent that Jane Dawson refers to a single movement she calls 'eco-nationalism' (1996).

Yet, following independence, the Baltic states were faced with overwhelming processes of social, political and economic change that challenged the Baltic societies existentially: for instance, questions regarding what it meant to be 'Estonian', 'Latvian' or 'Lithuanian': what it meant to be 'European'; and what it meant to be independent. Through the transitions to democratic-capitalism and the roads to memberships in the European Union (EU) and the North Atlantic Treaty Organization (NATO), questions over the environment were often lost amongst the struggles of transitioning societies to become less Soviet and more Baltic, if not more Nordic. Even with the focus of the European Commission on Estonia, Latvia and Lithuania in the accession process, the environment was not mentioned as an impediment to entry into the EU. On the surface, environmentalism seemed to disappear from public life in the Baltic states, despite having played an important role in the independence movements.

However, more recently environmental issues, or those pertaining to the environmental movement, have risen again as important items on the agenda. In Estonia, the planned 'Nordstream' natural gas pipeline connecting the Russian Federation and Germany has been stopped from going through Estonian waters based on the claim of possible ecological damage (Karm 2008). At the same time, the politics surrounding 'Nordstream' indicate that geopolitics matter more than ecological issues, at a time when Estonia continues to rely heavily on its oil-shale resources. In Latvia, the world's first ever Green prime minister was able to pull together a minority coalition to govern the country for a little less than a year in 2004. On reflection, former Green Party Prime Minister Indulis Emsis was not necessarily Green in the ecological sense, but rather owed his politics to other special interest agendas. Finally, in Lithuania, the late Soviet demands for the decommissioning of the Ignalina nuclear reactor succeeded in the post-Soviet era. However, recently with worries around energy security, the Lithuanian government has embarked on the development of a new reactor: Ignalina II.

Nevertheless, environmentalism carries on in the Baltic states. It no longer garners the attention or fosters the collective action that it once did, but the relationship with nature in the Baltic states continues to be an important part of 'nation' and 'self'. Nationalist politicians continue to use nature as a backdrop to their own political agendas. Non-governmental organizations (NGOs) continue to play an active role in conserving if not preserving the environment. Finally, with Europeanization, and some might argue globalization, attitudes towards the environment and the way in which Baltic societies are responding to ecological challenges are changing in terms of activism, policies, and international environment development funds. Yet, with a generally agreed threat of climate change, if not global warming, these low-lying littoral states have much to lose in a rapidly changing environment and as there is little doubt that the environment will once again return to the agenda as it once did at the rebirth of the three Baltic nations.

Environmental discourses

Before we come to contemporary environmentalism in the Baltic context, let us first cover what we mean by the environmental movement. The environmental movement in North America and Western Europe resulted from the protest movements of the

1960s and 1970s. There was a call not only to halt harm to the natural environment but also to reverse any previous harm. Several different discourses can be seen within the environmental movement in relation to what John S. Dryzek (1997, p. 12) refers to as industrialism, or the need for 'overarching commitment to growth in the quantity of goods and services produced and to the material well-being which that growth brings with it'. The focus on industrialism, rather than, say, capitalism, allows us to look at environmentalism across a series of ideologies, making the transition from socialism to capitalism more easily discussed in such a conceptual framework.

Dryzek identifies two discursive axes of the environmental movement. The first axis is the reformist-radical axis. *Reformist* discourses take the industrial environment as given and attempts to alter the negative externalities of industrialism through the system that relies on the industrial environment. *Radical* discourses challenge the nature of industrial societies and their inherent negative impact on the environment. Second, Dryzek identifies the prosaic-imaginative axis. *Prosaic* discourses argue that it is the excesses of the industrial society that leads to environmental degradation. On the other hand, *imaginative* discourses see environmental problems as opportunities which the industrial society can solve through an alteration of the social system. As the reformist-radical and prosaic-imaginative discursive axes overlap, one can identify four discourses (Figure 1).

'Environmental problem-solving' argues for the use of public policy to adjust the status quo to eliminate the worst excesses of industrialism. While seemingly the least contentious discourse, Dryzek (1997, p. 15) identifies the long-running debate between administrative and market-led mechanisms to counter high levels of air and water pollution, not to mention the more recent disputes over global warming. Second, 'survivalism' paints the world as heading towards cataclysm with ever increasing population growth and reducing resources. '[Survivalism] is radical because it seeks a wholesale redistribution of power within the industrial political economy and a wholesale reorientation away from perpetual economic growth' (1997, p. 14). Third, 'sustainability' argues for imaginative solutions within the current industrial society. This discourse is seen in contemporary European political arguments with the convergence interests of economic growth and environmental protection.

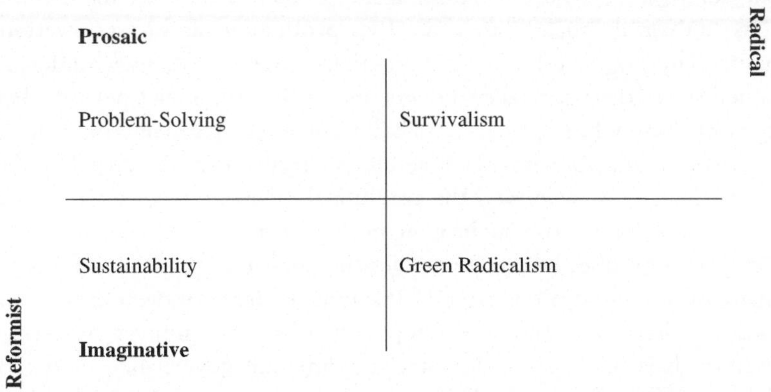

FIGURE 1 Environmental discourses (Dryzek 1997, pp. 12–14).

Finally, 'Green radicalism' seeks solutions to the damage done to the natural environment outside industrial society. Within this discourse, Dryzek (1997, p. 15) argues that 'green romantics disagree with green rationalists, proponents of the rights of individual creatures disagree with more holistic thinkers, and advocates of green lifestyles disagree with those who prefer to stress green politics'. Nevertheless, these disagreements center on a common radical vision of social transformation to deconstruct industrial society to suit the environment.

While the full discursive range can easily be seen in 'Western' politics and society, on which much of the environmental literature is centered, they can also be seen in the contemporary Baltic states. As we shall see, the stress put on political and economic transition has meant that the reformist discourses are far more powerful, even if not necessarily prevalent. However, as seen in this issue, radical discourses persist in the Baltic states, especially in connection to 'nation' and national identity. Eastern notions of nation have led to a strong emphasis on *heimat*, or caretaker societies, as opposed to Western notions of pristine nature. As we shall see, Baltic environmental discourses vary between the reformist political and economic agendas and the radical social agendas of nationalism. For this reason, the Baltic states are an interesting case for the study of environmentalism.

Contemporary environmentalism in the Baltic states

This collection sets out to uncover the state of environmentalism in the Baltic states since the end of the Soviet Union as well as to determine how Baltic societies and governments have responded to environmental issues in an era of Soviet pasts and European futures. Within this context, we set out to answer several questions. First, having achieved independence from the Soviet Union, what happened to those who acted and sympathized with the environmental movements of the late Soviet era? Timofei Agarin examines this question looking at environmentalists in the early stages of the post-Soviet phase in Estonia, Latvia, and Lithuania. Relying on the World Values Survey, Agarin takes a comparative look at those who saw themselves as environmentally conscious. He tests the concept put forward by Dawson of 'movement surrogacy', whereby nationalism projects itself initially by framing its demands as universal goals, such as the protection and conservation of the environment. The logic follows that if there was a genuine underpinning of environmentalism in the nationalist movements of the late Soviet period, there would be a remnant of those within post-Soviet Baltic societies that maintained and identified with the environmental movement. Overall, he argues that the social, political, and economic transitions overwhelmed the traditional environmentalists and forced them to focus on first-order needs, such as economic scarcity, job security, and national security. At the same time, he acknowledges the persistent presence of environmental NGOs, many of which began as part of the independence movements.

Richard Caddell takes the next step and asks what impact Sovietization and Europeanization have had on modes of environmental governance in the Baltic Sea Region. His essay focuses specifically on Estonia's entry into the EU and its effect on nature conservation in the country. Caddell looks at historical approaches to nature conservation in Estonia with a well-established platform from which to protect the environment. Furthermore, the author looks at the logic behind the Estonian government's approach to nature conservation in the post-Soviet years, where he finds that the state was keen to reverse the environmental damage of the Soviet period while at the same time being keen to adopt EU directives for environmental protection. Altogether, Caddell takes a legalistic approach to the difficulties in implementing and harmonizing nature conservation legislation in Estonia. The essay sheds light on the European challenge to environmental protection in the Baltic sea region, where historical closeness to nature has been an inherent part of nation-building since the nineteenth century.

In this context, two essays look at the relationship between environmentalism and nationalism in the Baltic states. David J. Galbreath and Daunis Auers look at environmentalism and nationalism in terms of how the Latvian Green Party has attempted to use 'Green' discourse to hide a nationalist political agenda. The essay challenges the 'Green' credentials of the Latvian Green Party and the world's first ever 'Green' prime minister. Additionally, Tove Malloy looks at minority environmentalism and eco-nationalism in the context of the late Soviet period. She looks at the connection between environmentalism and nationalism in a general sense, illustrating how Green political theorists argue that the two ideologies are distinctly separate in their ends if not in their means. Malloy sets out to see how minority environmentalism, the use of the environment as a way to frame minority grievances, interacts with nationalism. She argues that minority environmentalism and minority nationalism are not as antithetical as Green political theorists would argue. However, once minority nationalism becomes majority nationalism, the conflict between the universal principles of environmentalism and the local principles of sovereignty are ultimately diverging principles.

What impact has the environmental agenda had on contemporary Baltic politics? Galbreath and Auers go further in their discussion of the Latvian Green Party with the question 'green, brown, or black?' The essay looks at the different sides to the party that espouses environmental conservation, beats a nationalist drum, and is funded by the oil transit oligarch, Aivars Lembergs. Allan Sikk and Rune Holmgaard Andersen examine the often-found Western relationship between environmentalism and socialism in contemporary Estonian politics. They argue that unlike the post-modern Green parties in Western Europe, the Estonian Green Party is essentially 'ultra-modernist' in terms of its party platform, center-right politics, and electoral support. In the two essays, the authors illustrate how the environmental movement represents an alternative ecological discourse in Latvia and Estonia that obscures the existing genuine 'Green' movements in the Baltic Sea Region.

Finally, two essays address the question of what impact the post-Soviet environmental challenge has had on local approaches to environmental protection. Nora Mžavanadze analyses the relationship between local councils and NGOs and their effects on sustainable development programs. She argues that sustainable development is fundamentally tied to strong social capital more generally. In this light, Mžavanadze examines the failure of the Lithuanian National Strategy for Sustainable Development. She argues that the strategy was essentially professionalized to such an extent that it cut out public participation, which impaired its basic implementation.

Jussi S. Jauhiainen and Tarmo Pikner examine local, and in this case international, relationships over common environmental challenges. Jauhiainen and Pikner look at the changes surrounding the shared water infrastructure between the Estonian city of Narva and the Russian city of Ivangorod. The authors argue that changes in the water supply system had a substantial impact on local communities and increased the value of the basic commodity, tap water. In the case of the Narva–Ivangorod water network, a local problem of infrastructure was compounded by the geopolitics of the post-Soviet region. In this way, the collection ends with a reflection on the transborder nature of the environment, illustrating environmentalism's important contribution to contemporary Baltic studies.

Overall, the essays within the collection represent interesting comparative and case study approaches to contemporary environmentalism in the Baltic states. We give Estonia, Latvia and Lithuania in-depth country analyses, as seen in Caddell, Galbreath and Auers, Sikk and Andersen, Mžavanadze, and Jauhiainen and Pikner. We also provide comparative studies, including Agarin and Malloy. Together, these essays deliver three aspects of contemporary environmentalism in the Baltic states. First, environmentalism's relationship with nationalism and nation-building is complex in terms of how they weave together while ultimately diverging in their respective goals. Second, the environmental movement is likely to suffer during times of political and economic uncertainty as it did in the early years of the post-Soviet period. This conclusion bodes ill for contemporary environmental movements in the Baltic states during this time of global economic recession. Finally, environmentalism in the Baltic states illustrates an interesting crossroads between the radical discourses of Western European 'Greens' and NGOs with the reformist discourses found within the socio-political contexts specific to Estonia, Latvia and Lithuania. In total, the contributors to this collection provide a well-researched and insightful analysis on the nature of environmentalism in the Baltic states.

Acknowledgement

I would like to thank David J. Smith, Editor, and Sarah Lennon, Technical Editor, of the *Journal of Baltic Studies*, for their encouragement and constructive comments. The contributors also wish to acknowledge the excellent comments and suggestions from an anonymous referee.

References

Dawson, J. I. (1996) *Eco-Nationalism: Anti-Nuclear Activism and National Identity in Russia, Lithuania, and Ukraine* (London, Duke University Press).

Dryzek, J. S. (1997) *The Politics of the Earth: Environmental Discourses* (Oxford, Oxford University Press).

Karm, E. (2008) 'Environment and Energy: The Baltic Sea Gas Pipeline', *Journal of Baltic Studies*, 39, 2, pp. 99–122.

WHERE HAVE ALL THE ENVIRONMENTALISTS GONE? BALTIC GREENS IN THE MID-1990S

Timofey Agarin

In the wake of *Perestroika*, environmental issues stood out in the eastern Baltic Sea region as being behind the forces successfully mobilizing mass support for independence movements. At the time, these forces challenged the supremacy of the central apparatus to decide on issues of high concern to citizens, which allowed for an initial dialogue between the Soviet state and society. Arguably, mass movements in the then Soviet Baltic republics strongly opposed the extensive model of using natural resources and urged the authorities to account for the ecological repercussions of the Soviet path to economic modernization. However, the tide of environmental protest receded after the Baltic states gained independence from the Soviet Union (SU) and many issues which had mobilized citizens just months earlier lost their prominence altogether. By the mid-1990s, political, economic and social changes further curtailed civic engagement, including environmental initiatives, within the societies of the Baltic states.

This essay analyses the views of those Baltic residents who support parties with an environmental agenda and of those involved in 'Green' activism. The essay seeks to provide an explanation for an overall decline of popular interest in 'Green' politics and activism in the Baltic states by investigating the world views, individual environmental action, and support for environmental policies across the region. Relying on the data collected during the second round of the World Values Survey (WVS), I investigate whether Baltic respondents who support environmental parties and members of environmental groups display the individual attitudes typically deemed characteristic of Western European 'Greens'. In doing so, the essay primarily reflects on research findings on environmental attitudes outside the post-socialist world, which suggest that support for 'Green parties' goes hand in hand with an individual's environmental concerns. The essay also addresses findings from the study of Western European politics, which indicate that 'Green' voters and members of environmental organizations are distinguished by sociotropic values and hence support a 'Green' agenda and environmental policies.

There is a host of literature explaining the reasons for the highly successful mobilization of Soviet citizens into popular movements during the years of *Perestroika* (Auer 1998; Muiznieks 1990, 1995; Rinkevicius 2000; Schwartz 2005, 2006; Shtromas 1990). However, little research has been done to explain the sudden reduction in the number of environmental organizations following independence from the SU. This essay takes the debate over the Baltic 'Green' movements during the late socialist period as its starting point, but questions the salience of environmentalism in post-Soviet Baltic politics. While political parties running on the 'Green' platform did enter Baltic parliaments in the early 1990s, they slid into oblivion by the end of the decade, garnering only meager popular support and gradually disappearing from the political landscape as an independent parliamentary force.

In order to see where Baltic environmentalists have gone, I investigate the answers of some 3000 respondents from Estonia, Latvia and Lithuania collected in the 1996/1997 wave of the WVS. First, I discuss the attitudes of the electorate attracted to Baltic parties with 'Green' programs. This study demonstrates that the Green electorate in the Baltic states shows little appreciation for what is generally held to distinguish Western European 'Green' voters from the rest of their fellow voters. However, the essay does not argue that Baltic societies are disengaged from environmental activity. To verify whether or not this is the case, the second part of the essay focuses on the members of environmental organizations and discusses the structure of their beliefs. This part of the essay makes clear that while some people holding 'Green' views supported 'Green politics' in the mid-1990s, today's 'Green' voters are not the most environmentally minded among the respondents pooled in the WVS study. Does this scenario mean that 'Green' issues receded from the political agenda in Estonia, Latvia and Lithuania by the mid-1990s? Or does it mean that perhaps Baltic environmentalists are not so 'Green' after all? The essay provides a starting point for research into Baltic environmental values.

Are 'Green politics' relevant in the Baltic societies?

Most studies of environmental activities in today's post-Soviet Baltic states tend to underline that contemporary activities are but a shadow of the ecological mobilization of the 1980s (Auer 1998; Pae *et al.* 2005; Rinkevicius 2000; Schwartz 2007). In the late Soviet period, environmental protests mobilized around a quarter of the republics' residents in opposition to the development of oil shale mining in Estonia, the erection of new hydroelectric dams in Latvia, and nuclear power plant expansion in Lithuania. The majority of those who have analyzed ecological mobilization during this time argue that it was the 'national awakening' that provided the background for these activities. Indeed, since the *samizdat* article 'Internationalism or Russification' by Ukrainian Ivan Dziuba, environmental issues had been present in dissidents' calls for considering the ecological impact of socialism on the Soviet nationalities.

In discussing the rise of ethnonationalism throughout the SU, Dina Zisserman-Brodsky suggests that national movements representing dispersed ethnic groups disregarded environmental issues, 'while other groups . . . , generally regarded any unfavourable ecological situation in their region as the result of deliberate manipulation' (Zisserman-Brodsky 2003, p. 89). For example, the account of the national movement in Nagorno-Karabakh demonstrates that environmental deterioration was of high concern for the local populace, even if it only epitomized the ethnic grievances of the local Armenians (Geukjian 2007). In the Baltic republics too, numerous local groups were active in urban regeneration activities (Auer 1998), landscape maintenance (Schwartz 2007), and the monitoring of industrial organizations (Rinkevicius 2000). Although significant numbers of people were mobilized through these groups, they never framed their environmental grievances as issues of political concern. Thus, the local initiatives provided the emergent national movements with a unique opportunity structure for national mass mobilization.

Studying the late Soviet ecological movements, Jane Dawson suggests that Soviet popular movements were particularly successful where they used what she calls 'movement surrogacy' (Dawson 1996). In *Eco-nationalism*, she argues that perceptions of the harmful environmental impact of Soviet economic developments allowed for the expression of public discontent within the climate of *glasnost*'. Ecological problems were portrayed as yet another assault on local national identities, resulting in controversies that pitted the central authorities against republican leaderships, academics and environmental activists (Dawson 1996, pp. 35–41). She suggests that while ecological issues were high on the political agenda and rallied the population, they had little relevance to people's everyday life. Thus, the movements' ability to embrace a broad range of social issues was key to their success. Since many citizens across the SU felt particularly connected with their environment, popular movements could not help but embrace 'Green' slogans.

The 'Dawson hypothesis' was later supported by Rinkevicius. Investigating the ecological rhetoric of *Sąjūdis*, he suggests that it was nothing but a counterweight to the Soviet technocratic approach to ecological, economic, and, importantly, social issues (Rinkevicius 2000). Similarly, studies of mass ecological movements in Estonia and Latvia underline that their success was guaranteed by linking a national lifestyle and heritage to perceptions of landscape and nature (Kionka 1990; Schwartz 2006; Trapans 1991). In so far as little space was allocated to strictly environmental issues in the Baltic popular movements, 'movement surrogacy' provides the best explanation for their initial success. However, 'surrogacy' also provides an explanation for the movements' quick dismemberment after the restoration of independence.

Once the movements' major opponent, the Soviet regime, ceased to exist, they struggled to survive in a new political environment and fragmented into various groups with political, as well as non-political agendas (Grofman *et al*. 2000; Muiznieks 1995; Trapans 1991). This is what we observe across the former Soviet Union (FSU) and in the Baltic states, where *Sajūdis, Latvijas Tautas Fronte*, and *Eestimaa Rahvarinne* disintegrated into numerous political parties, pressure groups, and citizen initiatives in the run-up to the first post-Soviet elections (Clark 1998; Ishiyama 1993; Shtromas 1990). During this period, political elites began 'outsourcing' divisive issues to non-political groups, while at the same time profiling themselves with a distinguishable political agenda.

The fragmentation of interests on the political stage, as well as the priority given to economic development after the collapse of the Soviet redistribution mechanism, placed additional constraints on the environmental movements in the region. In particular, the costs of the post-Soviet economic recession were frequently connected to previous environmental activities, undermining the legitimacy of public participation in environmentally sensitive policy issues (Rinkevicius 2000). This logic explains the limited environmental advocacy of *Eesti Rohelised* (Estonian Greens, ER), *Latvijas Nacionāli Konservatīvā/Latvijas Zaļā Partija* (Latvian National Conservative Party and Latvian Green Party, LNK/LZP) and *Lietuvos Žalioji Partija* (Lithuanian Green Party, LZP). While the number of supporters for the 'Green' parties dwindled, one could still expect to see environmental activism prospering outside the domain of politics. However, environmental organizations remained small and marginal across the Baltic states. Does this mean that Baltic residents grew uninterested in environmental issues?

The research of environmental activism and the 'Green' vote in Western Europe has become well developed over the past decade. Particularly, the 'social capital' literature suggests that membership in environmental groups reflects the levels of general trust in a particular society rather than a connection with the perceptions of environmental deterioration, political opportunity structure, or social awareness of the problems (Sonderskov 2008; see also Mžavandadze 2009). However, the studies examining the downswing in environmental activism argue that this is not indicative of the pragmatic decisions of actors or their subjective cost-benefit calculations (Kousis 1999). Instead, scholars investigating the growth of environmental activism suggest that environmental action derives from an interplay between political, economic, social, as well as cultural factors (Frank *et al.* 2000). Some claim that an egalitarian social environment and democratic stability have a positive impact on the number of environmental groups, but not on the breadth of their engagement (Dalton 2005). Many of these studies echo Ronald Inglehart's claim of a global shift in values from materialist to post-materialist orientations (1990, 1997). Public awareness of the negative impact of human activity on the environment, Inglehart argues, leads individuals to consider bearing the individual costs of environmental protection. This includes greater public engagement with environmental groups, paying an environmental tax, and ultimately expanding the numbers of political actors engaging in 'Greening' the political agenda (Inglehart 1995).

While the majority of the literature on the basic environmental orientations of the Baltic populations has applied a discourse analytical approach to the study of 'Green' politics and environmental activists (Rinkevicius 2000; Schwartz 2007), empirical data exist to allow a comparison of voters' behavior and their life orientations. Among these, the New Baltic Barometer (NBB) prominently features as the best cross-country survey with a specific focus on the preferences of the Baltic populations (Rose 2005). However, data from the NBB do not allow a consecutive analysis of relations between the ecological attitudes of respondents, their political behavior and non-political activities. The World Values Survey (WVS) on the other hand, while providing an empirical base for cross-cultural comparison in values and orientation of the public, includes a range of indicators on environmental attitudes. These allow for a regression analysis on voting preferences and declared environmental activism. The object of the analysis – variable perceptions of nature, the environment, and human relations with these – additionally speaks in favor of using the WVS set.

The data collected in the WVS suggest the growing acceptance of environmental thinking in different societies across the world and the rise of support for environmental parties in Western Europe since the early 1980s (Inglehart 1997, p. 248). Although the explanatory power of the WVS model has been questioned, it does provide the best empirical base available to analyze the relation between social and political processes (Davis *et al.* 1999; Warwick 1998). Decisively the stability of the social and economic conditions is essential for sustainable social development and gradual political change, all of which induce cognitive independence and increase awareness of complex relations between humans and nature (Inglehart & Welzel 2005, p. 24).

In fact the analyses of the WVS surveys located 'Green' issues within the nexus of three mutually reinforcing factors: environmental awareness, environmentally friendly everyday activity, and preferences to address 'Green' issues through politics (Norris 2002; Norris & Inglehart 2004). The students of 'human development' generally interpret the diminishing propensity of the population to engage in environmental initiatives and vote for 'Green' parties as evidence of lower levels of human agency, which prevents individuals from embracing environmental ideals fully (Inglehart 1984, 1995; Klingemann 1999; Welzel *et al.* 2003). At the same time, more recent investigations suggest that environmental activism is an additional, rather than an independent indicator for a particular type of interpersonal relations (Dalton 2005; Inglehart & Welzel 2005; Sonderskov 2008). Whether this is the case across the Baltic societies should be left for the consideration of more skilled scholars. This essay's focus is on the environmental attitudes and related mindset of the 'Green' voters, as well as of those active in the environmental organizations throughout the Baltic region. My view of the individual values infers a bottom-up logic: individual attitudes influence preference formation, which in turn impacts group membership choice and voting patterns.

In the following part of the essay I investigate specific attitudes which distinguish those voting for 'Green' parties outside the post-socialist world, those who support an environmental agenda and those who favor changes in the human approach to the environment.

Investigating 'Green voters'

The research on supporters of 'Green politics' distinguishes two groups of voters: some respondents support 'Green' parties because they expect greater environmental accountability, while another group does so for non-environmental reasons (Neumayer 2004). Largely, these studies distinguish between respondent attitudes and actions (Rootes 1995). Thus, much of the research on environmental politics in Western Europe indicates that not all respondents voting 'Green' are

'Green'-minded, although most of them demonstrate a coherent set of environmental beliefs (Eckersley 2004; Rihoux & Rüdig 2006). The WVS includes contextual variables that offer a general picture of respondent attitudes and actions. These variables have been classified into sets, describing the respondents' demographic backgrounds and general views on the one hand; on the other hand individual environmental action and 'Green political thinking' were singled out (see appendix).

In order to provide a competent account of those respondents who indicated their preparedness to vote for 'Green parties', I investigate four hypotheses which result from observations of the Western European 'Green electorate'. Research on the 'Green vote' suggests that higher education, a greater degree of social security, such as stability of job and higher income, and the perception of oneself as belonging to the middle-class all positively influence the choice of the 'Green' political alternative (Burningham & Thrush 2003; Kreuzer 1990; Bennulf 1990; Muller-Rommel 1985). These findings inform my first hypothesis: socio-demographics matter.

Growing environmental activism and the prominent movement of parties running on environmental platforms in Western European politics over the past two decades highlight the qualities of their electorate, which tends to sympathize with leftist ideals (Poguntke 1993; Rihoux & Rüdig 2006; Rüdig 2006). More generally, of course, environmental activists attempt to ensure the general accessibility of environmental goods, additionally portrayed as not particularistic and thus inherently leftist. Although the ideals associated with the 'left' differ considerably across different societies, reliable markers distinguishing the 'left' from the 'right' include attitudes toward social change and private ownership (Bernt 1990; Faucher 1998). However, across the FSU respondents are said to be cautious in showing any compassion for 'leftist' ideals, which can easily be mistaken for sympathy with the former socialist regime (Evans & Whitefield 1998). By including questions on the endorsement of radical political change and the degree of satisfaction with the way the current political system works, I can balance the question of left-wing orientations with the post-socialist respondent who might see the state ownership of business through the lens of dialectical materialism. Together, the three questions allow for an uncontroversial 'pro-left' classification. This leads to the second hypothesis exhibited in Western European electoral behavior: an individual preference for ensuring the accessibility of public goods encourages a 'Green' vote.

Similarly, studies on the links between political behavior and individual worldviews suggest that 'all big things start small'. In the West, a potential 'Green' voter invests considerable time into 'everyday environmentalism', such as choosing environmentally friendly products, reducing water usage, and engaging with local environmental organizations and groups (Blake 1999; Phil 2003; Poortinga *et al.* 2004). 'Everyday environmentalists' opt for environmental action even in the situation where the costs for non-compliance with social expectations are easily avoided (Georg 1999; Lorek & Spangenberg 2001). Compared with other types of environmental activity, such as a contribution to local environmental organizations, patterns of consumption behavior can indicate an individual's awareness of personal environmental impact which orders his/her preferences accordingly. To what degree everyday environmental behavior impacts political preferences is tested under the hypothesis: environmental action is important. I expect to see evidence of 'Green' voters' more coherent environmental action patterns in their everyday choices, more so than in voters of other political parties.

Electoral studies suggest that 'Green' voters in Western European countries support taxes for environmental protection and favor ecological production even at higher prices while also preferring economic policies that do not affect the environment negatively (Ashworth *et al.* 2006; Kasa 2000; Thalmann 2004). Whether these observations are also relevant for the Baltic electorate is investigated in the final hypothesis: 'Green voters' support the political regulation of environmental issues. If Baltic 'Green' voters are comparable with Western European 'Greens', they will declare a personal preparedness to financially contribute to the regulation of economic development in relation to its impact on the natural environment through 'Green' taxes.

The WVS explores respondents' preferences for political parties by asking whom they would support in an election. Among the options available to respondents were *Eesti Rohelised*, *Latvijas Nacionāli Konservatīvā/Latvijas Zaļā Partija* and *Lietuvos Žalioji Partija*. However, as results indicate, only data from the 1996 wave for Latvia and Estonia can be considered in further analyses. In all other waves of the survey 'Green parties' were on the list of parties to choose from, but were not mentioned once by the respondents. Although the Lithuanian Greens were mentioned in the 1999 survey by four individuals as their first or second choice party, including these in the further analyses reduces the overall predictability of the results due to a very low sample. For the survey of 1999 ($N = 3{,}023$), the relation of 'non-Green' to 'Green' voters is below 1%, disallowing any comparison between these groups from a statistical point of view (see Table 1). Hence, only data from the Latvian and Estonian samples of 1996 will be considered to investigate my hypotheses. The valid sample remains at 217, accounting for 9.8% of the entire Estonian and Latvian samples ($N = 2{,}221$).

In order to assess whether socio-demographics impact individual preferences for *Eesti Rohelised* and *Latvijas Nacionāli Konservatīvā/Latvijas Zaļā Partija* five items have been included into the regression analysis which account for merely 2.1% variance of those voting Green. Only the younger age of a respondent impacts his/her preference for a 'Green' party, while all other items do not qualify the voter to be more likely to prefer an environmental party at an election. Neither gender, nor employment status, nor the perceived belonging to a particular social class indicates any correspondence with preference for a 'Green' party in Estonia or Latvia. These findings indicate that, contrary to the first hypothesis, there is no specific socio-demographic background of

TABLE 1 'Green' vote across Baltic states, second or first choice

	Yes		No		All
	N	%	*N*	%	*N*
Estonia (1996)	64	6.3	957	93.7	1021
Latvia (1996)	153	12.8	1047	87.2	1200
Lithuania (1999)	4	0.04	1014	99.96	1018
N	221		3018		3239

the voters of ER and LNK/ZP. Interestingly, although the level of education attained does not translate into support for the 'Green' political blocks, there is some relationship between a higher-level education among the younger respondents and propensity to vote 'Green'. The variance in the sample therefore appears to be explained almost entirely by the preference of the more educated younger Latvian and Estonian voters for the environmental parties.

Compared to socio-demographic variables, the items reflecting the general worldviews of the respondents voting 'Green' provide a somewhat more comprehensive explanation of the results. Around 3.4% of the entire variance in the sample could be accounted for by investigating ideological belief structures. However, whether the worldview behind the 'Green' vote can be described as oriented towards securing a common good over particularistic profits, as one would expect from the studies of Western European societies, is highly questionable. Among other indicators of the worldviews, the left–right polarization appears to bear the greatest explanatory power for the variance at a significant level. Counter to expectations, a majority of 'Green' voters in Estonia and Latvia identify themselves as sitting on the 'right' side of the ideological scale. This observation is reinforced by the fact that the second strongly expressed item in the pool – preference for governmental ownership of business – negatively predicts the 'Green' vote. This finding allows me to classify Baltic 'Green' voters as those who have right-wing preferences and particularistic notions of profit. Other items included in the model are of no statistical significance for the differences between 'Green' and other voters and have little impact on the overall variance, showing negligible Beta coefficients.

While the data do not support the second hypothesis that Estonian and Latvian 'Green voters' share leftist orientations, seen as so crucial for identifying West European 'Greens', the causality does not necessarily work in the opposite direction. The lack of support for 'Green politics' specifically, and the absence of a left-wing orientation generally, does not imply that individuals are not environmentally minded (Neumayer 2004). As Kollmuss and Agyeman (2002) argue, some basic tenets of environmentally friendly behavior have entered the commonly accepted norms of conduct and hence could be treated as a separate marker of environmental attitudes. These also have a high probability of influencing preferences for environmental activism and the 'Green vote' (Steel 1996). Following hypothesis three, patterns of individual environmental action do not suggest any distinguishable preferences with the voters of ER or LNK/ZP. Attending a meeting or petitioning for environmental protection, as well as choosing household products that one thinks are better for the environment, appears to have meager significance when it comes to distinguishing 'Green voters' from the voters of other political parties, explaining only 2.5% of the entire variance in the sample. These two items, however, influence the change in the overall variance considerably ($\beta = 1.094$ and $\beta = 0.740$, respectively). The declared choice of environmentally friendly products also has some positive impact on electoral preference for ER or LNK/ZP, while membership of an environmental organization is likely to correlate with an electoral preference for 'Green' parties. Specifically, it indicates that political parties catering to the ideals of environmentally minded voters closely collaborate with the organizations concerned with environmental issues. This finding suggests that ER and LNK/LZP collect some support from environmentally minded voters while at the same time not attracting all potential sympathizers of the 'Green' movement.

In this light, it is essential to investigate the relationship between individual-level attitudes toward social reality and a preparedness for collective action in the form of electoral behavior. The studies of Western European 'Green' voters have previously suggested that members of these groups are distinctly different from those professing environmental concerns in the surveys (Grob 1995). Personal preparedness to bear the cost of reducing the adverse effects of economic developments, should the 'Green' party pursue environmentally friendly policies upon entering office, would be a good indicator of coherent environmental beliefs.

Critically, my last hypothesis taps both dimensions of an individual's 'Green attitudes' and personal commitment to 'Greening' the socio-economic process through political regulations. Unfortunately, the WVS sample does not provide any evidence in support of the hypothesis that voters of 'Green parties' are prepared to contribute personally to 'Green taxes'.

Indicators showing support for the environmental regulation of business appear to bear strong explanatory power for the differences between 'Green' and 'non-Green' respondents, accounting for 2.9% of the entire variance. Declared voters of the ER and LNK/ZP appear to approach political processes from an angle radically different from their Western European 'Green' counterparts. Out of the four items within the regression analyses, only one demonstrates no correlation with the preference for a 'Green' party: the preparedness to favor environmentally friendly produce at a higher price. This could be indicative of the economic difficulties in the mid-1990s and the overall socio-economic insecurities of Baltic residents in the initial years of independence (see Rinkevicius 2000). This socio-economic context illustrates macrosocial factors that diminish the preference for environmental choices overall (Boratav *et al.* 1996). Unlike their Western European counterparts, Baltic respondents voting 'Green' have great support for the view that human beings should harness nature for their own ends. This difference is further substantiated by negative attitudes expressed towards increasing environmental taxes as well as general support for economic growth despite possible damage to the environment.

Altogether, these indicators demonstrate a clear negative correlation between Western concepts of sustainability and the Estonian and Latvian views of 'Green' politics. The results lead to a conclusion that 'Green' parties in Estonia and Latvia are electorally successful in spite of rather than because of their environmental agenda. Indeed, the attitudes of their electorate toward environmental sustainability and state-sponsored environmental policies indicate the opposite – those voting for Baltic 'Green' parties hold views that favor economic development even at the cost of environmental decline. Overall, the Baltic 'Green voter' is unlikely to hold the views of Western European 'Green' voters. In Latvia and Estonia, a 'Green' party voter is a 'rightist conservative' who disapproves of political intervention even at the cost of long-term degradation of the environment and welcomes the appropriation of nature as a human resource. The opinions of Estonian and Latvian respondents supporting ER and LNK/ZP lead to a conclusion which would follow the observations of Soviet environmental legacies in the Russian Federation. For instance, Jonathan Oldfield (2001) argues that Russian citizens are inclined to see nature not as an asset to be preserved, but rather as a resource for economic development too abstract to be granted individual time, finance or action.

In effect, we can conclude that the two parties under review do not represent all the votes of those residents who share environmentally friendly views. Instead, the voters seem to assess the importance of other issues on the agendas of these parties and only marginally buy into the 'Green' allure of LNK/ZP and ER. The attitudes of the 'Green' electorate suggest that many 'Green issues' are outsourced elsewhere, in what amounts to an echo of the late-socialist 'movement surrogacy'. This leads us logically towards an analysis of environmental organizations, as the most obvious 'containers' of environmentally minded individuals.

Who are the Baltic environmental activists?

While voters for 'Green' parties in the Baltic states are few in number, this does not mean that environmental concerns are absent from these societies. Given the initial strength of the environmental movement in the Baltic states, one might expect to find at least some 'Green minds' among those active in environmental organizations, if not among those who vote for ER and LNK/ZP. If the tacit implication of 'movement surrogacy' is true, than we will see that the 'Green' issues were 'outsourced' to non-political organizations such as environmental movements. In this case, a clear correspondence will be visible between environmental attitudes and the beliefs of members of Baltic environmental organizations. Also, I will be able to estimate the differences in orientations between the members of the environmental organizations and those claiming to support the 'Green' parties.

In the following I identify the attitudinal differences between those who belong to these organizations and those who do not. Second, I consider the differences within the former group, highlighting the aspects of value orientation distinguishing the active and non-active members of the environmental groups in Estonia, Latvia, and Lithuania. In assuming that the membership in these organizations is voluntary, even passive support indicates an individual's support for environmentalist causes. This assumption is further specified by the observation made in the previous part of the essay, namely that not all members of environmental organizations are likely to cast their ballot for a 'Green' party. In fact, the majority of members of the environmental organizations chose a different party to represent their political interests (see Table 2). Considering that there is some overlap between 'Green' party supporters and

TABLE 2 Membership in environmental organization and potential 'Green' vote

	Active member	Inactive member	Not a member	*N*
Will vote for 'Green' party either as first or second choice	7	8	202	217
No	11	44	2955	3010
N	18	52	3157	3227

environmental activists, some of those sampled in the earlier section will be a part of this analysis.

In the second wave of the WVS, respondents were asked to identify whether they were members of environmental organizations, and if they were, whether they were active or non-active (see Table 3). Overall only 70 individuals indicated that they were active or inactive members of environmental organizations, which accounts for 2.2% of the entire sample of 3230 surveyed across the three Baltic states. Needless to say,

TABLE 3 Membership in environmental organization

	Active member	Inactive member	*N*
Estonia (1996)	4	20	24
Latvia (1996)	10	14	24
Lithuania (1997)	4	18	22
N	18	52	70

TABLE 4 Results of regression analyses

	Green vote	Organizational membership	Active membership
Socio-demographics			
Age	−0.157**	0.007	−0.135
Male gender	0.202	−0.741**	−0.508
Higher education	0.214*	0.335	−0.122
Subjective class	−0.095	−0.102	0.138
Employed	−0.015	0.036	0.194
R square	***0.021***	***0.021***	***0.080***
Leftist orientation			
Left leaning	0.160**	0.124	−0.228
Radical social change	0.059	−0.479	−1.289*
Public ownership	−0.086**	−0.044	−0.075
Satisfaction with democracy	−0.156	−0.157	−0.213
R square	***0.034***	***0.020***	***0.153***
Environmental action			
Environmental products	0.522*	0.740**	−0.358
Reduce water use	0.291	0.170	−0.260
Recycled	#	−0.221	0.571
Attend meeting, petitioned	1.094***	#	#
Contributed to environmental organizations	0.774	#	#
R square	***0.025***	***0.015***	***0.159***
Political regulation			
Tax increase	−0.609***	0.712*	−0.838
Endorse higher price	0.178	−0.160	0.355
Support environ protection	−0.389**	−0.487*	0.143
Human coexists with nature	−0.590*	−0.062	−19.884
R square	***0.029***	***0.028***	***0.089***

Note: $^{*}p<0.05$; $^{**}p<0.01$; $^{***}p<0.001$. #item not used in regression.

this data set is a good candidate for a small-*N* analysis, which has particular limitations to its interpretation. One of the central inferences in this respect requires an *ad hoc* assumption on the causality of relations in the logistic regression (Ebbinghaus 2005; Mahoney 2000). With the relations inferred as going from environmental attitudes towards a respondent's engagement in environmental action, I expect that non-members of the environmental organizations will be less likely to demonstrate positive attitudes towards the environment and undertake individual actions. While the *N* is small, the result of the logistic regression accounting for up to 2.2% of the variance between the test groups could be considered solid. However, unless each set of independent variables explains the same or greater levels of variance in the sample, its impact on the dependent variable, in this case membership of organizations, should be considered spurious (Leiberson 1991, 1993).

Remarkably, the regression analysis of the differences in attitudes between members of environmental organizations and non-members is hardly any different from what has been previously observed with the voters of the 'Green' parties in Latvia and Estonia. Once again, socio-demographic variables appear to have little relevance for explaining the membership of environmental groups, accounting for 2.1% of the variance. Female respondents appear to be more active in environmental groups than males and, to the predicament of the 'Green' vote, leftist political views do not have any relevance for membership in environmental organizations. Here again, those individuals who identify themselves as holding views on the 'right' side of the political scale are more likely to be engaged in environmental organizations. This set explains only 2.0% of the variance. Neither is there much evidence that members of environmental organizations espouse individual environmental action, with the items on the scale explaining a mere 1.5% of the alternations in the sample. Within this set of items, the individual preference to choose environmentally friendly household products positively and significantly impacts upon the correlation ($\beta = 0.740^{**}$). Put simply, this means that not even members of environmental organizations have distinguishable environmental action patterns and are solely distinguished by preference for environmentally friendly products. Hardly surprisingly, the attitudes of members of environmental organizations towards the political regulation of ecological issues vary only marginally at the level of 2.8%. As with those who vote for 'Green' parties, members of environmental associations neither favor an increase in environmental taxation nor oppose economic growth at the expense of ecological deterioration. In other words, among the four models investigated, only the set on political intervention robustly reflects the differences between members and non-members of environmental organizations. At the same time, membership in environmental organizations spuriously correlates with an exploitative approach to nature and consistent support for the state withdrawing from environmental protection.

In my final attempt to establish any similarities between environmentally minded respondents in the Baltic states and those in Western Europe, I turn to the views of those who declare themselves to be members of environmental organizations. I compare the responses of 70 individuals, among whom 18 have declared themselves to be active, and 52 non-active members of associations. However, although both the socio-demographic variables and the attitudes towards political regulation of environmental issues present some division between the groups of active and non-active members, explaining the variance of 8.0% and 8.9% respectively, none of the items introduced into the analytical model appears to be statistically significance in this context. Two of the hypotheses – environmentally minded individuals have leftist political orientations and individual environmental action is important – address the variance in the pool of answers at the relatively high levels of 15.9% and 15.3% respectively.

The variance in active/non-active membership is best accounted for by attitudes towards environmental action, of which recycling can be discerned as particularly important for the active members. All other items, including the choice of environmentally friendly products or being aware of water usage, have no significance in this context whatsoever. What is more interesting is that active/non-active members of environmental organizations have differing views on how society should be changed. This item has the highest statistical significance in the model addressing leftist political orientations. Active members of environmental groups hold the view that social change should not be gradual, but radical in nature. Whereas Western European environmental activists frequently pursue change by means of radical action (DeLuca 1999; Rootes 2003), we cannot assume a similar sentiment within the group of Baltic respondents. Instead, one could assert more cautiously that environmental activism in the Baltic states is unlikely to produce conservationist approaches to nature and resources. However, it is not clear whether opposing conservationism in this case implies more active social regulation of environmental issues, or its opposite: the extensive use of nature for the purposes of economic regeneration.

In this context, I suggest that in the absence of environmentally friendly behavior patterns even among those respondents declaring active membership in the environmental groups, we should explore the collinear effects of 'Green' activism. These might include privileged access to economic and ecological resources, which are distributed among those activists, or that the activities help them accumulate political and social capital within the networks of environmental organizations. All these factors may prompt individuals to participate in 'Green' groups without incumbent changes in their behavioral patterns or attitudes. The analyses clearly suggest that even those individuals who declare an engagement with environmental organizations lack a coherently structured set of environmental beliefs.

The investigation of the views of WVS respondents active in Baltic environmental organizations does not suggest that there is any potential for an increase in the 'Green' vote across the Baltic states. This finding, however, does not preclude the potential for them to vote for the 'Green' parties in the region, which, as I established in the second part of the essay, are no more 'Green' than the environmental activists themselves. The WVS data therefore indirectly support Dawson's 'movement surrogacy' hypothesis and call for further investigation into environmental attitudes in the Baltic societies. Additionally, the data suggest that the structure of the environmental beliefs of the Baltic 'Greens' is radically different from their Western European counterparts. If anything, there is little reason for cooperation between the Western European 'Greens' and the Baltic environmentalists.

Discussion

The analyses presented in this essay have investigated the beliefs and attitudes among WVS respondents in the Baltic states who expressed support for 'Green' parties and are enrolled in environmental organizations. The results indicate that these attitudes of the 'Green' electorate and members of environmental organizations are not consistent with what has been previously observed among environmentally minded individuals in the Western European context.

This brings the success of the Baltic environmental movement's struggle against the Soviet regime to the fore: why were they successful and what were the reasons for the popular mobilization of their members? Clearly, political and economic conditions have changed dramatically since the popular mobilization of the 1980s and the time when the WVS survey was taking place. Some scholars argue that the difficulties of the 'triple transition' prompted Baltic residents to relocate their support from long-term aspirations toward more short-term benefits (Rinkevicius 2000). Ecological activities thus gave way to more immediate concerns, such as political and economic security and stability.

Although the background conditions for political support and organizational membership did change between the late 1980s and mid-1990s, it is acknowledged that individual values and priorities take much longer to modify. Thus, while the reasons for support of 'Green' issues in the region changed, the aptitude for environmental concern in the Baltic societies should have remained much the same. While values explaining preferences for environmental action, the choice of 'Green' parties, and membership in environmental organizations within West European societies are clearly evident, the timeframe since the demise of the SU could not produce the attitudinal change explaining the lack of environmental concern among the Baltic publics by the mid-1990s. If one takes individual-level attitudes as determinants of political agendas and patterns of social activism, the continuing downswing of any environmental voice in Baltic politics since the mid-1990s turns out to be expected.

The negligible numbers of 'Green'-minded individuals even among the members of environmental organizations suggest that there was perhaps never a great concern about the state of the environment throughout the region. From this point of view the central question of my essay – where have all the 'Greens' gone? – arguably gives rise to a more significant question: were there ever any environmentalists in the first place? Needless to say, according to the data on attitudes and preference of the members of these organizations, I do not sustain the claim that the WVS surveys collected the most environmentally friendly activists of the popular movements.

Reconstructing the individual motivation to back 'Green' parties in the region, the essay has provided support for the 'movement surrogacy' hypothesis. Clearly, the environment was not among the top concerns of the Baltic 'Greens' despite being integral to the late Soviet popular political mobilization and the subsequent success of the popular movements in the late 1980s. Although some were willing to support 'Green' parties in the mid-1990s, they were unlikely to vote for the environmental agenda of LNK/ZP and ER, and instead preferred other issues on their programs. This deflection of Green support is evidenced by the disappearance of the 'Green' vote from Baltic WVS respondents during subsequent years that fall outside the scope of this study. Given the fact that the voters of the Baltic 'Green parties' are not the most environmentally minded among the electorate it is no surprise that there has been further vote dispersion away from the LNK/ZP and ER since the mid-1990s. Unless these parties add additional issues to appeal to their previous voters, they look set to disappear from the political stage. The discussion of the political career of the LNK/ZP by Galbreath and Auers (2009) additionally makes clear that pragmatic, rather than environmental, issues facilitated this party's enduring presence in Latvian politics.

Of particular interest is the finding that both 'Green' voters and environmental activists in the Baltic states emphasize particularistic interests and underline their self-classification in terms of the political 'right'. This would indirectly support Jane Dawson's argument on the 'movement surrogacy' during *Perestroika*. As with many other scholars, Dawson argues that nationalist slogans dominated in the Baltic mass movements that used ecological platforms as a smoke screen to disguise other goals. The exposition to the Soviet regime, more frequently associated with the 'left' than with the 'right', could provide a background for individual disapproval of state intervention in the range of issues, including environmental regulation.

The very same experience, however, prompts individuals to expect social protection in the situation of structural disadvantage. Individual preference for particularistic interests, however contradictory, has been previously noted to be prone to exploitation in the Eastern, as well as the Western, European party systems (Marks *et al*. 2006). This argument is central for the explanation of the resurgence of populist parties across Western Europe (Lloyd 2003), and was deployed to address the ideological shift in the European radical left after the collapse of socialism (March & Mudde 2005). Thus, as the data indicate, if environmental movements attract individuals in pursuit of particularistic interests, they naturally feature as containers of the populist vote. On the broader scale, the data make clear that various particularistic interests such as 'right' or 'nationalist' converge in the Baltic environmentalist movements under the label of 'Green'. Although the data are too spurious to suggest that the local 'Greens' surrogate ecological concerns with nationalism, they clearly embrace the political views of the 'right'. Particularly in the face of the ongoing economic crisis, the stance of both ER and LNK/ZP on stricter immigration criteria and societal integration allows multiple parallels with the rhetoric of the new populist parties in Western Europe.

Conclusion

This essay investigated the attitudes of Baltic WVS respondents towards the environment and sought to uncover connections between individual attitudes, membership of environmental groups and preferences for 'Green' parties. Instead, my analyses suggest that there is a fundamental mismatch between the attitudes of the Baltic 'Greens' and those in the West. Additionally, I was able to demonstrate that the lack of consistent perspectives on human–environment relations within the Baltic societies explains the receding numbers of activists in environmental organizations.

As is clear from this analysis of member preferences of environmental organizations, lack of interest in environmental activity, personal disengagement from the practices regulating impact on the environment, and the emphasis on economic priorities over ecological sustainability all have an adverse effect on environmental activism in Estonia, Latvia and Lithuania. Clearly, this requires qualitative investigation on the future of the human–environment interface in the Baltic states.

While 'Green' politicians and environmentalists pay lip-service to environmental issues, it would appear that both are in fact driven primarily by a desire to accumulate other, more tangible assets and gain greater access to resources. Possibly, the more optimistic conclusion might be that Baltic environmentalists are not active in organizations that state 'Green' goals explicitly. In that case further research is needed to assess why previously sound environmental voices have opted for silent accommodation with current realities. Whatever the answer to these questions, 'Green' parties and environmental activism have a decreasing presence within the Baltic states.

References

Ashworth, J., Geys, B. & Heyndels, B. (2006) 'Determinants of Tax Innovation: The Case of Environmental Taxes in Flemish Municipalities', *European Journal of Political Economy*, 22, 1.

Auer, M. R. (1998) 'Environmentalism and Estonia's Independence Movement', *Nationalities Papers*, 26, 4.

Bernt, A. (1990) 'Green Politics: A Norwegian Experience', *Scandinavian Political Studies*, 13, 2.

Blake, J. (1999) 'Overcoming the "Value-Action Gap" in Environmental Policy: Tensions between National Policy and Local Experience', *Local Environment*, 4, 3.

Boratav, K., Türel, O. & Yeldan, E. (1996) 'Dilemmas of Structural Adjustment and Environmental Policies under Instability: Post-1980 Turkey', *World Development*, 24, 2.

Burningham, K. & Thrush, D. (2003) 'Experiencing Environmental Inequality: The Everyday Concerns of Disadvantaged Groups', *Housing Studies*, 18, 4.

Clark, T. D. (1998) 'The 1996 Elections to the Lithuanian Seimas and Their Aftermath', *Journal of Baltic Studies*, 29, 2.

Dalton, R. J. (2005) 'The Greening of the Globe? Cross-National Levels of Environmental Group Membership', *Environmental politics*, 14, 4.

Davis, D. W., Dowley, K. M. & Silver, B. D. (1999) 'Postmaterialism in World Societies: Is It Really a Value Dimension?', *American Journal of Political Science*, 43, 3.

Dawson, J. I. (1996) *Eco-Nationalism: Anti-Nuclear Activism and National Identity in Russia, Lithuania, and Ukraine* (London, Duke University Press).

Ebbinghaus, B. (2005) 'When Less Is More: Selection Problems in Large-N and Small-N Cross-National Comparisons', *International Sociology*, 20, 2.

Eckersley, R. (2004) *The Green State: Rethinking Democracy and Sovereignty* (Cambridge, MA, MIT Press).

Evans, G. & Whitefield, S. (1998) 'The Evolution of Left and Right in Post-Soviet Russia', *Europe-Asia Studies*, 50, 6.

Faucher, F. (1998) 'Is There Hope for the French Ecology Movement?', *Environmental Politics*, 7, 3.

Frank, D., Hironska, A. & Schofter, A. (2000) 'Environmentalism as Global Institution', *American Sociological Review*, 65, 1.

Galbreath, D. G. & Auers, D. (2009) 'Green, Black and Brown: Uncovering Latvia's Environmental Politics', *Journal of Baltic Studies*, 40, 3.

Georg, S. (1999) 'The Social Shaping of Household Consumption', *Ecological Economics*, 28, 3.

Geukjian, O. (2007) 'The Politicization of the Environmental Issue in Armenia and Nagorno-Karabakh's Nationalist Movement in the South Caucasus 1985–1991', *Nationalities Papers*, 35, 2.

Grob, A. (1995) 'A Structural Model of Environmental Attitudes and Behaviour', *Journal of Environmental Psychology*, 15, 3.

Grofman, B., Mikkel, E. & Taagepera, R. (2000) 'Fission and Fusion of Parties in Estonia, 1987–1999', *Journal of Baltic Studies*, 31, 4.

Inglehart, R. (1984) 'The Changing Structure of Political Cleavages in Western Societies', in Dalton, R.J., Flanagan, S.C. & Beck, P.A. (eds) (1984) *Electoral Change in Advanced Industrial Democracies* (Princeton, Princeton University Press).

Inglehart, R. (1990) *Culture Shift in Advanced Industrial Societies* (Princeton, NJ, Princeton University Press).

Inglehart, R. (1995) 'Public Support for Environmental Protection', *PS: Political Science & Politics*, 28, 1.

Inglehart, R. (1997) *Modernization and Postmodernization. Cultural, Economic, and Political Change in 43 Societies* (Princeton, NJ, Princeton University Press).

Inglehart, R. & Welzel, C. (2005) *Modernization, Cultural Change, and Democracy: The Human Development Sequence* (Cambridge, Cambridge University Press).

Ishiyama, J. T. (1993) 'Founding Elections and the Development of Transitional Parties: The Cases of Estonia and Latvia, 1990–1992', *Communist and Post-Communist Studies*, 26, 3.

Kasa, S. (2000) 'Policy Networks as Barriers to Green Tax Reform: The Case of CO_2-taxes in Norway', *Environmental Politics*, 9, 4.

Kionka, R. (1990) 'The Estonian Citizens' Committee: An Opposition Movement of a Different Complexion', *RFE/RL Report on the USSR*, February 9.

Klingemann, H.-D. (1999) 'Mapping Political Support in the 1990s: A Global Analysis', in Norris, P. (ed.) (1999) *Critical Citizens. Global Support for Democratic Governance* (Oxford, Oxford University Press).

Kollmuss, A. & Agyeman, J. (2002) 'Mind the Gap: Why Do People Act Environmentally and What Are the Barriers to Pro-Environmental Behavior?', *Environmental Education Research*, 8, 3.

Kousis, M. (1999) 'Sustaining Local Environmental Mobilisations: Groups, Actions and Claims in Southern Europe', *Environmental Politics*, 8, 1.

Kreuzer, M. (1990) 'New Politics: Just Post-Materialist? The Case of the Austrian and Swiss Greens', *West European Politics*, 13, 1.

Leiberson, S. (1991) 'Small N's and Big Conclusions: An Examination of the Reasoning in Comparative Studies Based on a Small Number of Cases', *Social Forces*, 70, 2.

Leiberson, S. (1993) 'More on the Uneasy Case for Using Mill-Type Methods in Small-N Comparative Studies', *Social Forces*, 72, 4.

Lloyd, J. (2003) 'The Closing of the European Gates? The New Populist Parties of Europe', *The Political Quarterly*, 74, Supplement 1.

Lorek, S. & Spangenberg, J. H. (2001) 'Indicators for Environmentally Sustainable Household Consumption', *International Journal of Sustainable Development*, 4, 1.

Mahoney, J. (2000) 'Strategies of Causal Inference in Small-N Analysis', *Sociological Methods & Research*, 28, 4.

March, L. & Mudde, C. (2005) 'What's Left of the Radical Left? The European Radical Left after 1989: Decline and Mutation', *Comparative European Politics*, 3, 1.

Marks, G., Hooghe, L., Nelson, M. & Edwards, E. (2006) 'Party Competition and European Integration in the East and West: Different Structure, Same Causality', *Comparative Political Studies*, 39, 2.

Muiznieks, N. R. (1990) 'The Latvian Popular Front and Ethnic Relations', *RFE/RL Report on the USSR*, October 20.

Muiznieks, N. R. (1995) 'The Influence of the Baltic Popular Movements on the Process of Soviet Disintegration', *Europe-Asia Studies*, 47, 1.

Muller-Rommel, F. (1985) 'Social Movements and the Greens: New Internal Politics in Germany', *European Journal of Political Research*, 13, 1.

Mžavanadze, N. (2009) 'Sustainable Development in Lithuania: Between the Government Agenda and the Undiscovered Civil Society', *Journal of Baltic Studies*, 40, 3.

Neumayer, E. (2004) 'The Environment, Left-Wing Political Orientation and Ecological Economics', *Ecological Economics*, 51, 3–4.

Norris, P. (2002) *Democratic Phoenix. Reinventing Political Activism* (Cambridge, Cambridge University Press).

Norris, P. & Inglehart, R. (2004) *Sacred and Secular: Religion and Politics Worldwide* (Cambridge, Cambridge University Press).

Oldfield, J. D. (2001) 'Russia, Systemic Transformation and the Concept of Sustainable Development', *Environmental Politics*, 10, 3.

Oldfield, J. D. (2005) *Russian Nature: Exploring the Environmental Consequences of Societal Change* (Aldershot, UK & Burlington, VT, Ashgate Publishing Limited).

Pae, T., Luud, A. & Sepp, M. (2005) 'Artificial Mountains in North-East Estonia: Monumental Dumps of Ash and Semi-Coke', *Oil Shale*, 22, 3.

Phil, M. (2003) 'Embodying the Environment in Everyday Life Practices', *The Sociological Review*, 51, 1.

Poguntke, T. (1993) 'Goodbye to Movement Politics? Organisational Adaptation of the German Green Party', *Environmental Politics*, 2, 3.

Poortinga, W., Steg, L. & Vlek, C. (2004) 'Values, Environmental Concern, and Environmental Behavior: A Study into Household Energy Use', *Environment and Behavior*, 36, 1.

Rihoux, B. & Rüdig, W. (2006) 'Analyzing Greens in Power: Setting the Agenda', *European Journal of Political Research*, 45, s1–s33.

Rinkevicius, L. (2000) 'Transitional Economies', *Environmental Politics*, 9, 1.

Rootes, C. (1995) 'Environmental Consciousness, Institutional Structures and Political Competition in the Formation and Development of Green Parties', in Richardson, D. & Rootes, C. (eds) (1995) *The Green Challenge* (London, Routledge).

Rose, R. (2005) *New Baltic Barometer VI: A Post-Enlargement Survey* (Glasgow, Centre for the Study of Public Policy).

Rüdig, W. (2006) 'Is Government Good for Greens? Comparing the Electoral Effects of Government Participation in Western and East-Central Europe', *European Journal of Political Research*, 45, s127–s154.

Schwartz, K. Z. S. (2006) *Nature and National Identity after Communism: Globalizing the Ethnoscape* (Pittsburgh, University of Pittsburgh Press).

Schwartz, K. Z. S. (2007) '"The Occupation of Beauty": Imagining Nature and Nation in Latvia', *East European Politics and Societies*, 21, 2.

Shtromas, A. (1990) 'How Political Are the Social Movements in the Baltic Republics?', *Nationalities Papers*, 18, 2.

Sonderskov, K. M. (2008) 'Environmental Group Membership, Collective Action and Generalised Trust', *Environmental Politics*, 17, 1.

Steel, B. S. (1996) 'Thinking Globally and Acting Locally?: Environmental Attitudes, Behaviour and Activism', *Journal of Environmental Management*, 47, 1.

Thalmann, P. (2004) 'The Public Acceptance of Green Taxes: 2 Million Voters Express Their Opinion', *Public Choice*, 119, 1.

Trapans, J.A. (ed.) (1991) *Towards Independence: The Baltic Popular Movements* (Boulder, Westview Press).

Warwick, P. V. (1998) 'Disputed Cause, Disputed Effect: The Postmaterialist Thesis Re-Examined', *Public Opinion Quarterly*, 62, 4.

Welzel, C., Inglehart, R. & Klingemann, H.-D. (2003) 'The Theory of Human Development: A Cross-Cultural Analysis', *European Journal of Political Research*, 42, 3.

Zisserman-Brodsky, D. (2003) *Constructing Ethnopolitics in the Soviet Union: Samizdat, Deprivation, and the Rise of Ethnic Nationalism* (Basingstoke, Palgrave).

APPENDIX Variable labels

Variable label	Description	Range
Independent variables		
Green vote	If there were a national election tomorrow, for which party on this list would you vote? Just call out the number on this card. If DON'T KNOW: Which party appeals to you most?	*Eesti Rohelised, Latvijas Nacionāli Konservatīvā/Latvijas Zaļā Partija*, or *Lietuvos Žalioji Partija*=1 Else=0
Organizational membership	Belong to conservation, environmental, ecology, or animal rights groups	1=yes 0=no
Active membership	Active/Inactive membership of environmental organization	1=active 0=inactive
Dependent variables		
Age	Age collapsed	1=less then 18 years old 2=18–29 years 3=30–39 years 4=40–49 years 5=50–59 years 6=more then 60 years
Education	What is your highest level of education?	1=Primary school 2=Incomplete secondary 3=Secondary (without certificate) 4=Vocational secondary 5=Secondary (with certificate) 6=Technical college 7=Incomplete higher 8=University degree
Gender		1=male 2=female
Employment status	Are you employed now or not?	1=Yes 0=No
Subjective class		
Leftist orientations	In political matters, people talk of 'the left' and 'the right'. How would you place your views on this scale, generally speaking?	1=left 10=right

Social change	On this card are three basic kinds of attitudes concerning the society we live in. Please choose the one which best describes your own opinion.	0 = 'Society must be radically changed' 1 = 'Society must be gradually improved by reforms'
Ownership	Private ownership of business should be increased versus Government ownership of business should be increased	1 = 'Private ownership of business should be increased' 10 = 'Government ownership of business should be increased'
Satisfaction with democracy	On the whole are you very satisfied, rather satisfied, not very satisfied or not at all satisfied with the way democracy is developing in our country?	1 = 'Very satisfied' 2 = 'Rather satisfied' 3 = 'Not very satisfied' 4 = 'Not at all satisfied'
Environmental action	Which, if any, of these things have you done in the last 12 months, out of concern for the environment? Have you chosen household products that you think are better for the environment?	0 = 'Have not' 1 = 'Have done'
Water use	Have you tried to reduce water consumption for environmental reasons?	0 = 'Have not' 1 = 'Have done'
Meetings, petitions	Have you attended a meeting or signed a letter or petition aimed at protecting the environment?	0 = 'Have not' 1 = 'Have done'
Contribute to environmental organizations	Have you contributed to an environmental organization?	0 = 'Have not' 1 = 'Have done'
Tax increase	I would agree to an increase in taxes if the extra money were used to prevent environmental pollution	1 = 'Agree' 0 = 'Disagree'
Higher price	I would buy things at a 20% higher price if it helped to protect the environment	1 = 'Agree' 0 = 'Disagree'
State intervention	Here are two statements people sometimes make when discussing the environment and economic growth. Which of them comes closer to your own point of view? A. Protecting the environment should be given priority, even if it causes slower economic growth and some loss of jobs B. Economic growth and creating jobs should be the top priority, even if the environment suffers to some extent	1 = 'Protecting environment' 0 = 'Economy growth and creating jobs'
Human coexist with nature	For each of the following pairs of statements, please tell me which one comes closest to your own views: A. Human beings should master nature B. Human beings should coexist with nature	1 = 'Human beings should coexist with nature' 0 = 'Human beings should master nature'

NATURE CONSERVATION IN ESTONIA: FROM SOVIET UNION TO EUROPEAN UNION

Richard Caddell

The dissolution of the USSR and the transition to democracy of its former constituent republics has revealed a deeply alarming legacy of environmental degradation and damage across the land, air and sea space of the former Soviet Union. The Chernobyl disaster and the startling retreat of the Aral Sea provide us with the most graphic illustration of the ecological excesses arising from Soviet economic policies. These policies gave little consideration to long-term environmental impacts or mitigation strategies. In many instances, the newly independent countries of the former Soviet Union have inherited substantial environmental problems requiring extensive and expensive remediation activities.

In the case of Estonia, the challenge of remediation has been considered 'comparatively more formidable than in other Baltic Republics of Latvia and Lithuania' (Auer 2005, p. 117). The ravages of oil shale extraction and phosphate mining (Taagepera 1989, p. 165), as well as the widespread dumping of military waste, often containing toxic and radioactive contaminants, threatened both the ecological balance of the country and the long-term health of the population. The restoration of a clean and safe environment was thus a high priority in post-Soviet Estonia (Auer & Raukas 2002, p. 679). In the period between the collapse of the Soviet Union and Estonia's accession to the European Union, the country's environment underwent an astonishing transformation. This period thus presents a fascinating case study of environmental management, encompassing a sea change in legislative cultures, administrative structures and government perceptions of ecological concerns, driven to a considerable extent by the need to implement EU standards in order to meet accession targets.

This essay provides a case study of this process as applied to a distinct sectoral issue of environmental governance – nature conservation. This focus is neither accidental nor arbitrary. While the protest movement of the 'phosphate spring' is widely considered to have advanced Estonian environmental consciousness regarding ecologically deficient Soviet policies, a vibrant and proactive network of advocates and activists for nature conservation can be traced back to the mid-nineteenth century. Such considerations also occupied a small but significant position on the legislative agenda of the inter-war Estonian Republic prior to the Soviet annexation. Moreover, nature conservation commitments constitute some of the most exacting standards incumbent upon EU Member States, thereby providing an illuminating example of the regulatory contortions required of a candidate country prior to accession.

This essay will therefore advance three main arguments. First, Estonia has proved to be a nature conservation actor of understated significance during the twentieth century, having established a promising tradition of protective measures before annexation by the Soviet Union had a primarily negative effect upon such developments. Second, EU membership is highly significant for Estonian conservation efforts, as the Habitats Directive has exerted a strong influence on contemporary nature conservation law. Nevertheless, while Estonia has developed an impressive network of protected areas, the Habitats Directive will establish searching practical commitments in respect of these sites, which may prove more challenging than the original good progress towards the establishment of the Natura 2000 network.[1] Third, EU membership will also affect Estonian commitments under multilateral conservation regimes, as the European Commission (hereinafter 'the Commission') places an increasing degree of importance upon a common position within international biodiversity fora, which will further affect the direction of national policies.

To this end, this essay will first trace the emergence of national policies from antiquity to the present day and critically analyze such developments. It will then discuss the broad nature-conservation provisions incumbent upon Estonia under EU law and evaluate the future development of domestic policies that will be required in order to meet these demanding and well-policed obligations. Finally, consideration will be given to some of the broad commitments established under international law and the increasing influence of the Commission in shaping national approaches.

Nature conservation legislation in Estonia: historical perspectives and current structures

Estonia has a long history of promoting legislation and other legal measures to address nature conservation concerns, with records of environmental provisions dating back to antiquity. This section will provide an overview of the development of modern nature conservation law and an outline of key current legislation. In this regard, a series of broad periods of legislative activity may be identified. First, a number of environmental decrees and ordinances applicable to present day Estonian territory, introduced by a series of conquering regional powers, provided a primitive legal framework to address nature conservation. Subsequently, the roots of the modern system may be traced back to the mid-nineteenth century with the establishment of pioneering scholarly societies and the emergence of the first modern protected areas – some of which continue to survive and flourish under current legislation.

The twentieth century, however, saw the enactment of a substantial body of legislation, commencing with the first explicit nature conservation laws introduced in the 1930s by the newly independent Republic of Estonia. Annexation by the USSR in 1940 led to a new – and in many respects rather unfavorable – regime of ecological management, in which the Estonian SSR nevertheless proved to be a proactive exponent of innovative environmental legislation. Latterly, the collapse of the Soviet Union and the subsequent (re)independence of Estonia has ushered in a new era of nature conservation law, in which the need to address the Soviet environmental legacy and to traverse the emerging path to EU membership has further exerted a substantial, and broadly positive, influence.

Early developments

The earliest surviving documented legislation to address nature conservation in Estonia was enacted in 1297 by the Danish King Erik Menved, to prohibit logging and tree felling on a series of islands off the coast of modern-day Tallinn (Sepp *et al.* 1999, p. 163). Prior to this, nature conservation was seemingly the product of early superstitions in relation to sacred trees and woodlands, in a manner broadly analogous to the Celtic druidic tradition. Indeed, to a considerable extent, subsequent legislation was centered on the need to regulate and protect forestry resources. Swedish rule in the sixteenth century saw the emergence of further restrictions on tree-felling activity, with regulations adopted in 1647 and 1664 establishing the strict protection of certain species of trees deemed especially important for food and forest management, culminating in the establishment of a large protected forest zone along the Baltic coast (Sepp *et al.* 1999, p. 163). Despite these developments, however, until the nineteenth century most such initiatives were largely *ad hoc* in nature and were concerned predominantly with practical land management to address the demands of a primarily agrarian population.

The Tsarist era

The development of wider environmental measures was facilitated by the concerted study of local ecosystems by scientists and naturalists among a populace of increased formal education. In this respect, a watershed moment was the founding in 1853 of the Estonian Naturalist Society (*Eesti Looduseuurijate Selts*, ELS), which continues to occupy a position of national ecological significance in the present day. Initially a constituent part of the wider Livonian Society of General Utility and Economy, the increasingly broad-ranging focus of the ELS soon met with disapproval within the mother organization, which was primarily concerned with agricultural issues. From 1868, the ELS became an independent organization closely affiliated with the University of Tartu, through which it primarily collated and expanded upon *ad hoc* research work conducted by a growing number of amateur scientists and naturalists active throughout Estonia. Accordingly, as a more intellectual appreciation of the natural environment began to flourish and find a centralized outlet, the development of environmental policy was primarily driven by individual endeavor and participation in special-interest groups, and motivated essentially by the need to protect national monuments.

Nevertheless, ecological concerns began to foster an ethos of nature conservation at many levels of society beyond the enthusiastic amateur and professional academician. Indeed, although motivated primarily by property-use considerations, it was considered that '[l]and owners had the best opportunity for conserving nature, and provisions pertaining to preservation of trees and boulders were even added to rental contracts' (Klein 2008, p. 6). By the early twentieth century, a series of studies conducted by the ELS emphasized the need to establish protected areas or nature sanctuaries to conserve key species of flora and fauna within Estonia. In 1910, such calls led to the establishment of the first formal protected area in Estonian territory, with the designation of a bird sanctuary in the Vaika Islands off the western coast of Saaremaa.

The (first) Republic of Estonia

Following the establishment of the Vaika Islands Bird Sanctuary, the ELS developed a series of biodiversity-related expert groups, inaugurating a distinct section for Nature Conservation in 1920, followed by specialist sections on Ornithology (1921) and Botany (1928), which further expanded the knowledge base in these specific areas. With Estonia having secured independence following a series of ongoing conflicts, and with a legislature keen to exercise its newly acquired law-making powers, the prospects for specific national environmental legislation were highly promising. In 1924 the first official nature reserves were designated, with protected areas established in Kastre-Peravalla and Harilaid, as well as for the Abruka forest and in respect of yew habitat in Tahkuna (Sepp *et al.* 1999, pp. 163–4).

In 1929 a first draft nature conservation law was developed. However, it failed to receive either debate or legislative approval as it was technically an *ultra vires* measure under contemporary administrative law on the basis that prospective statutes that gave rise to excessive public costs were automatically precluded from government consideration. Following extensive redrafting, on 17 December 1935 the Nature Conservation Act formally entered into force as the first specific biodiversity law enacted by an independent Estonian authority. The 1935 Act focused primarily upon *in situ* conservation of particular species and habitats and, in keeping with traditional environmental concerns until that point, placed concerted emphasis upon the preservation and maintenance of national monuments. The Act was augmented in 1936 with the establishment of a Nature Conservation Register listing protected trees, rivers and reserves, which was periodically updated in line with an increasing number of designations.

The 1935 Act was subsequently repealed by the Nature Conservation Act 1938, which expanded the coverage of the earlier statute to include public parks and other common areas, which had not originally been caught by the definitions previously advanced. The designation of protected sites and areas continued at a considerable pace: by the point at which Estonia ultimately became embroiled in the Second World War, some 549 entries had been made on the National Conservation Register, including the listing of 26 different species of plants and 47 separate nature reserves (Klein 2008, p. 7).

The Estonian SSR

As noted above, Soviet policies are widely – and not erroneously – blamed for leaving a legacy of severe environmental degradation, which the post-communist states have been forced to expend considerable financial resources and practical endeavor in addressing. The environmental harm wreaked by military pollution in Estonia is well documented and, by common consensus, remediation of such damage remains a long-term project. Nevertheless, it would be incorrect to suggest that the admirable focus upon the development of norms to address nature conservation concerns in Estonia evaporated during the Soviet period. Indeed, the Estonian SSR may realistically be considered to have been a significant actor within the Soviet Union in developing biodiversity legislation. Notwithstanding the loss of *de jure* legislative independence upon annexation by the USSR, considerable work continued throughout the Soviet era to advance the cause of nature conservation in Estonia.

When Estonia became subsumed into the USSR it inherited the Soviet system of nature conservation. In a number of respects – at least superficially – the Soviet conservation program shared similarities with the pre-existing Estonian regime. The Soviet system was somewhat derivative of the environmental program previously operational in Tsarist Russia, which had established a network of protected areas, known as *zapovedniki* (Pryde 1986, p. 356). A Soviet *zapovednik*, however, had little obvious contemporary Western equivalent. As one leading commentator on nature management in the USSR has remarked, 'it is not possible to make a precise analogy between them and any particular type of American land management institution; some most closely resemble the American concept of a national monument, others are more like wildlife refuges, and a few resemble our national parks' (Pryde 1972, p. 45). Nature conservation, as an ideal seemingly supported enthusiastically by Lenin (Pryde 1972, p. 50), initially held a strong profile in the early years of the Soviet Union. Moreover, on paper at least, the protection afforded to such areas was sweeping, with *zapovedniki* subject to legislative ring-fencing as 'forever withdrawn from economic utilisation for scientific-research and cultural-education purposes' (Pryde 1972, p. 212). Indeed, such sentiments are arguably far stronger than those established under the present version of the EU Habitats Directive,[2] which offers considerable scope for economic activity in protected areas (de Sadeleer 2007, p. 36).

Nevertheless, Soviet *zapovednik* practice rarely lived up to its legislative rhetoric. In a totalitarian system in which grandiose industrial and agricultural targets were frequently prescribed, the ecological integrity of such sites was routinely compromised for 'economic utilisation', while *zapovedniki* could also disappear with the same dizzying velocity at which they were often created. Indeed, 'reorganizations' of the *zapovedniki* framework were neither infrequent nor especially predictable; in 1951, for instance, a reorganization program saw swingeing cuts to network coverage, with the result that 'over two-thirds of the preserves and seven-eighths of the protected area was abolished' (Pryde 1972, p. 50). Further revisions to the Soviet network, if not necessarily as draconian as the cull of 1951, frequently involved changes to site designation and a downgrading of what was often an illusory protected status.

Against this background, within the Estonian SSR the ELS continued to advocate further protection for the environment, to seemingly little avail. Nevertheless, in the years immediately following the death of Stalin – who had proclaimed the ability to 'master the forces of nature in the USSR' (Pryde 1972, p. 9) – a series of promising developments emerged in the Estonian SSR and, indeed, throughout the Soviet Union as a whole. In 1955, a Commission for Nature Conservation was created in Tartu in order to administer the Republic's conservation program. Ultimately, however, there was little in the way of comprehensive legislation in place within the USSR to advance these responsibilities. In this respect, the Estonian SSR acted as a pioneer and, 'as a result of the groundwork laid by the Commission for Nature Conservation' (Klein 2008, p. 8), a Nature Conservation Act was enacted in 1957 and became the first such statute of its type established in the Soviet Union.

The introduction of the 1957 Act had two immediate consequences. In the first instance, the other Soviet Republics swiftly emulated the Estonian SSR, and between 1958 and 1963 similar statutes were introduced across the USSR to address more localized environmental concerns. Second, and more importantly for the Estonian SSR, a legislative impetus was provided to formally establish national *zapovedniki*, with the pre-1940 protected areas lying outside the Soviet system and therefore of questionable legal status to Moscow officials – if not necessarily to local operators. Accordingly, in 1957 the first four Estonian *zapovedniki* were established, in relation to Matsalu, to protect resting Baltic waterfowl; Nigula, to promote further research into peat bog ecology; the original protected area within the Vaika Islands, to protect nesting seabirds; and Viadumae, to protect fauna indigenous to Saaremaa (Pryde 1972, pp. 202–4). Further innovation occurred in 1971 with the designation of a *zapovednik* in respect of Lahemaa, which was the first protected area to be established in line with international, as opposed to Soviet, criteria and principles (Klein 2008, p. 8). Moreover, in 1981 the Lahemaa *zapovednik* became the first national park in Estonia to be afforded a targeted development plan to further advance its conservation prospects (Klein 2008, p. 8).

In spite of these innovative developments, it appears that the conservatory effect of these measures was highly limited in practice. Indeed, writing in 1972, Erik Kumari, then Chair of the Commission for Nature Conservation, noted that the Estonian legislation was struggling to ensure that the consumption of natural resources was conducted in a manner that would safeguard the needs of future generations (Kumari 1972, p. 230). In an outspoken criticism of the status quo, Kumari argued that '[t]he law on nature conservation adopted in Estonia in 1957 already fails to answer its purpose under the present conditions . . . We, too, need a new, up-to-date, law on nature conservation, on a new, wider, foundation' (Kumari 1972, p. 231). However, Kumari's appeal fell on largely deaf ears, and no major amendments of the 1957 Act were forthcoming during the Soviet era. Instead, conservation efforts mainly revolved around compiling a series of national Red Data Books as an inventory of key threatened species and the further designation of protected areas within Estonian territory.

If conservation efforts within the Estonian SSR proved less effective than under the previously independent regime, this was primarily due to the constraints and inefficiencies of the Soviet environmental system imposed upon Estonia following annexation in 1940. Indeed, a number of factors may be identified as inhibiting the further development and optimal operation of nature conservation laws throughout the USSR that strongly justified Kumari's call for sweeping legislative reform.

Firstly, as alluded to above, prior to 1957 there was little coordinated environmental legislation within the Soviet Union to prescribe clear targets and operational requirements. Combined with a certain institutional ambivalence towards environmental considerations when confronted with the demands of economic development and major socialist projects, *zapovedniki* thereby tended to be established and relinquished in an uncoordinated, arbitrary and *ad hoc* manner, to the detriment of nature conservation efforts across the Soviet Union.

Moreover, and perhaps this was the most serious failing of the Soviet nature conservation system, there were significant operational failings in the practical administration of the *zapovedniki* network, as well as other key policy areas. This unsatisfactory position was caused by an excessive degree of centralization and the virtual preclusion of regulatory improvisation. There was little devolution of responsibility for the various *zapovedniki*; instead, while protected areas could be proposed by an individual SSR, key decisions affecting such areas were rendered through executive orders from Moscow, which inevitably led to wholesale inefficiency in addressing the conservation needs of protected sites. As Misiunas and Taagepera note, '[b]ecause of Soviet emphasis on the "branch principle", most of the polluting industry was under the control of central ministries in Moscow, while pollution damage was most evident to territorial agencies' (Misiunas & Taagepera 2006, p. 293). Likewise, the operational emphasis upon centralized policy and output demands often led to the imposition of unsustainable harvesting quotas for key natural resources. For instance, such requirements are widely considered to have significantly impeded woodland governance in a country with a long history of effective forest management (Kallas 2002, p. 325).

In addition, certain sectoral policies advanced on a central level inflicted considerable damage upon the Estonian environment. While the difficulties raised by phosphate mining and other mineral extraction operations are well documented, the policy of agricultural collectivization had a strong adverse impact on a national landscape rich in meadows and open grazing areas. Such problems were exacerbated by a strong military presence in Estonia, which closed key areas of natural habitats to both professional and amateur naturalists, leading to considerable data shortages in relation to important species and ecosystems. The need to address this data deficiency was accordingly established as a key priority in the post-Soviet environmental reconstruction (Klein 2008, p. 9). Furthermore, the wildlife legislation established within the USSR suffered markedly from drafting difficulties in comparison to the present day. In the first instance, such laws were heavily anthropocentric in nature with little concept of nature protection for its own intrinsic value. Moreover, as Pryde observed in the 1980s '[t]he right to wildlife to exist for its own sake appears to depend on "good behaviour"' (Pryde 1986, p. 353). In many cases, the laws in question also focused solely on animal species, thereby neglecting flora and vegetation and creating additional practical difficulties in the conservation of endangered species (Kalda *et al.* 1992, p. 201).

Notwithstanding these adverse impacts of the Soviet system upon Estonian land use and natural resource policies at the time, the wholesale inefficiency of the Moscow-based system also presented certain unintentional ecological benefits inasmuch as certain areas of the USSR were spared the intensive development activities that they might otherwise have experienced during this period. Indeed, in many Western European countries at this time, a more efficient degree of industrial planning led to the wholesale overconsumption of natural resources, in a manner vividly illustrated by Garrett Hardin's classic model of the 'tragedy of the commons' (Hardin 1968, p. 1243). Thus, in addition to the positive aspects of the pioneering nature conservation policies developed within the Estonian SSR during this period, there was also a limited degree of 'protection by omission', with the heavy military presence in Estonian territory and disjointed industrial policies of the Soviet Union serving to prevent further environmental degradation.

Yet the extent to which the failure of the Soviet Union to fully exploit development opportunities in the Estonian SSR may have ultimately contributed to the preservation of natural resources remains largely an exercise in ironic conjecture. The development activities that did occur during this period still left substantial adverse environmental impacts. In order to remedy these and improve environmental standards, the restored Estonian Republic has been obliged to bring wholesale changes to both the legal and administrative culture.

The post-Soviet era

Generally speaking, environmental policies in the years following independence have pursued two broad aims. In the first instance, in the immediate post-Soviet period, there was a need to prioritize environmental legislation to remedy the most pressing difficulties posed by the USSR system, while salvaging and consolidating the more positive aspects of the previous regime. Latterly, as EU candidacy developed from remote possibility to practical reality, there was a need to aggregate national legislation as far as possible with European standards as a prerequisite to accession. Since Estonia regained independence, legislative developments in the specific context of nature conservation have also followed this general pattern.

In the immediate post-Soviet years, practical conservation policies were, to a considerable extent, centered on two key areas of activity. In the first instance, there was a pressing need to address the national knowledge base on vulnerable species and habitats, given that, in many respects, data collection had stagnated under Soviet control. During the Soviet era, large swathes of Estonian territory had been placed under military jurisdiction, which generally afforded minimal access to naturalists (Klein 2008, p. 9). Moreover, collectivization had placed highly unfamiliar demands upon the countryside, the ecological cost of which had to be swiftly assessed. Accordingly, between 1998 and 2000, a major research census was undertaken of natural habitats and species throughout Estonia in order to ascertain a strong scientific basis for further conservation activity (Saarniit 2004, pp. 17–18) in the manner previously advocated by Kumari during the Soviet period (Kumari 1972, pp. 230–1).

Secondly, and conversely, within the territories that *were* open to public examination, a vibrant and active scientific community had continued to pursue research efforts (Kumari 1971, p. 317). Thus, without losing sight of the substantial challenges of documenting ecosystems within newly demilitarized zones, it was also necessary in the short term to ensure that the status quo in existing areas was preserved. In this regard, particular attention was paid to ensuring that the protected status of existing *zapovedniki*, established under (and in some cases predating) the Soviet system was not diminished in a rapidly expanding free-market economy (Klein 2008, p. 9).

As far as legislative developments during this period are concerned, perhaps the most significant event was the adoption in 1992 of the Constitution of the Republic of Estonia. In this respect, two particular provisions of the Constitution may be identified as establishing a basis for the advancement of further measures to address environmental concerns. First, on a general level, Article 5 states that the 'natural wealth and resources are national riches which shall be used sustainably', a provision that provides a legal basis for the elaboration of further norms to advance sustainable development. Indeed, in 2005 the *Riigikogu* adopted a National Strategy on Sustainable Development (Riigikogu 2005, p. 1), within which one of the four key goals to attain by 2030 was identified as 'Ecological Balance'. Perhaps more significant in this regard is the other leading environmental clause of the Constitution, Article 53, which is adopted under a Chapter on 'Fundamental Rights, Freedoms and Duties'. Article 53 states that'[e]veryone has a duty to preserve the human and natural environment and to compensate for damage caused to the environment by him or her. The procedure for compensation shall be provided by law'.

To date, however, these constitutional provisions have inspired limited legislative and judicial activity. As Andresson and Kolk have observed, '[i]n the lack of court practice the possible uses of these provisions remain unclear' (Andresson & Kolk 2003, p. 140). The opaque quality of these provisions has also created difficulties of interpretation and scope (Andresson & Kolk 2003, p. 149). Indeed, in the words of Saarniit, 'despite the clear constitutional basis of the environmental law, the legal framework is still fragmented. Many new laws have been adopted, but in the case when new laws do not exist, the laws adopted during the Soviet period regulate the situation' (Saarniit 2004, p. 14).

One of the first new laws to be enacted was the Law on Protected Natural Objects 1994 (LPNO),[3] which formed the primary legal basis for Estonian nature conservation in the immediate post-Soviet period. Echoing the pre-Soviet view of protected areas, the LPNO designated a series of sites as qualifying for enhanced legal protection on the basis of varied criteria, including ecological value and compliance with international commitments (Article 3 – see below). Nevertheless, a number of shortcomings within the Act soon became apparent – not least in relation to implementation, where definitional deficiencies and an excessive reliance on framework legislation served to undermine the practical operation of the LPNO regime. Indeed, in many respects a Soviet-style hangover lingered in the choice of management techniques, and the Act was unable to function without the further adoption of *ad hoc* ministerial regulations to give effect to its conservatory provisions. Given that such secondary legislation was essentially competing for drafting attention within a highly pressurized legislative environment during the post-communist transition years, and thereby accorded a comparatively low priority, 'implementation in many cases [remained] impossible' (Veinla 1996, p. 50).

Notwithstanding these deficiencies in practical operation, attention soon turned towards the development of further policies to advance the conservation value of protected sites further. To this end, in the late 1990s a prevailing view began to emerge that 'nature conservation cannot be viewed as a separate field of activity, but instead, that activities necessary for protection and preservation of natural assets must be integrated with all other facets of life in a state, and done in the most sustainable and balanced manner' (Klein 2008, p. 10). Such a position corresponds strongly with the 'Integration Principle' advanced under Article 6 of the current EU Treaty,[4] and now accordingly established as a general principle of EU law (Dhondt 2003, pp. 15–110). In the light of these developments, the LPNO was amended in 2001 in order to establish a distinct program of subsidies for the monitoring of meadows, grasslands and forests, with such subsidies paid 'with the purpose of preserving nature through practical human activities' (Klein 2008, p. 10).

This emerging emphasis on the broad alignment of national environmental principles with those of Community law was not coincidental, as Estonia had by now formally applied for membership of the EU. Indeed, as Auer observes, such a process was largely replicated throughout the candidate countries of Central and Eastern Europe, where 'acceding to European environmental norms became nothing less than an obsession in the early years of the new century. For most EU candidates, transposing EU environmental legislation and developing credible domestic environmental legislation were one and the same' (Auer 2005, p. 175). Yet, despite this essential process of alignment with the *acquis*, and the almost evangelical belief in the virtues of EU environmental law expressed by a number of candidate countries, a sense of disillusionment has been observed in certain newly inaugurated Member States with regard to particular policies (Gille 2005, p. 14). Many such studies remain essentially *ad hoc* in nature and often touch upon areas of environmental policy that have been heavily criticized by environmental lawyers for their inefficiency and ineffectiveness (Davies 2003, pp. 217–66). As noted below, the Habitats Directive in particular is subject to considerable drafting weaknesses and, while it is considered that the incorporation of stronger EU standards in respect of both conservation and institutional demands has had an ultimately positive effect in Estonia, it should be remembered that the osmotic absorption of these norms as an essential candidacy requirement also involved the incorporation of the weaker aspects of this legislation, which might – or might not – have otherwise been avoided by the national legislature.

In any event, the process of aggregation of nature conservation norms in Estonia was not entirely painless, with the European Commission expressing considerable reservations concerning the ability of the LPNO to meet the demands of the environmental *acquis*[5] as an essential precursor for EU membership. Indeed, in 1997, reporting on the initial Estonian application, the Commission stated that 'although sectoral litigation has been passed, applicable standards and regulations are often still those of the Soviet Union. The new legislation tends to be of a framework nature, needing to be fleshed out with standards, implementation measures and administrative structures' (European Commission 1997, p. 90). While such legislative obstacles were considered to be surmountable in the medium term, some anxiety was expressed over the numbers of trained personnel able to address these shortcomings in practice. According to the Commission, '[a]n important bottleneck which needs to be addressed is the lack of human resources with expertise to conduct the approximation process' (European Commission 1997, p. 90).

The Commission continued to monitor national endeavors to meet EU membership criteria and observed a continual – if somewhat gradual – process of improvement in relation to nature conservation issues. In 2000, an important milestone was reached with the adoption of a long-term plan to implement the requirements of the Habitats Directive in relation to the designation of protected areas (European Commission 2001, p. 1747). By 2002, however, considerable shortcomings remained, not least the need to further align national legislation with that of the EU, as well as 'to strengthen administrative capacity, at both ministerial and local government level. Estonia also needs to continue to improve enforcement of environmental legislation' (European Commission 2002, p. 1403). On the eve of enlargement, the Commission considered that national discrepancies in comparison with EU standards were rectifiable in the relatively short period before accession, but that a considerable revision of the LPNO was required in order to attain a satisfactory level of harmonization. Moreover, such legislative developments would need to be accompanied by further efforts 'to finalize the list of proposed sites of Community interest, to designate special protection areas by accession, and to ensure that relevant protection measures are applied by accession' (European Commission 2003, p. 675).

In line with these sentiments, the LPNO was ultimately repealed and replaced with a further, more comprehensive, statute that more closely resembled the Habitats Directive in both purpose and terminology. The Nature Conservation Act 2004 was passed on 21 April 2004 and entered into force on 1 May, concurrent with Estonian membership of the EU. In this regard, the Act represents the primary legal basis through which the considerable demands of EU nature conservation law and policy are to be discharged in Estonia.

Nature conservation in Estonia in the EU era

The primary norms of EU nature conservation law are the Wild Birds Directive[6] and the Habitats Directive,[7] which prescribe a series of obligations to be transposed by the Member States. In recent years, these measures have been supplemented by additional provisions, most notably in the form of the European Community Biodiversity Strategy[8] and its accompanying Biodiversity Action Plans, which prescribe a broad outline of aspirational objectives. Furthermore, in 2002 the EC adopted its Sixth Environmental Action Programme,[9] which placed a strong emphasis upon the need to halt biodiversity loss throughout the Community by 2010 in line with commitments incumbent upon the EU through participation in the Convention on Biological Diversity, as outlined in the section 'EU nature conservation law' below. Accordingly, this section will first outline in broad terms the key obligations imposed by the two cornerstone nature conservation directives, before evaluating current practice of the Nature Conservation Act in the light of these guiding commitments and ascertaining potential areas of difficulty in the transposition of these provisions by the Estonian authorities.

EU nature conservation law

The Wild Birds Directive is one of the most long-standing and entrenched provisions of EU environmental law and it must also be considered to be one of the leading and most effective regional instruments to regulate anthropogenic activities affecting avian fauna. Since its adoption in 1979 the Wild Birds Directive has played a key role in the conservation of birds. While it is very difficult to assess objectively the 'effectiveness' of nature conservation laws, recent scientific research suggests that the directive has impacted favorably on the conservation status of the majority of species addressed under its auspices (Donald *et al.* 2007, p. 810).

Article 5 of the Wild Birds Directive requires Member States to establish a general system of protection for avian species addressed by the directive. In particular, this commitment involves taking the requisite measures to maintain the population of such species (Article 2), with an emphasis upon preserving, maintaining or re-establishing areas of habitat primarily through the creation of protected areas (Article 3). The main operative provision of the Wild Birds Directive in relation to habitats may be considered to be Article 4, which mandates that the species listed in Annex I of the directive are to be subject to 'special conservation measures' in respect of habitats so as to ensure the survival and reproduction of the species in their areas of distribution. Accordingly, Member States must classify the 'most suitable territories in number and size as special protection areas [SPAs], taking into account the marine and land areas subject to the directive'.[10]

Likewise, the Habitats Directive is perhaps the best-known provision of EU environmental law and continues to form the primary basis for regulating most species of flora and fauna, both at a Community level and within individual Member States. The primary aims and objectives of the Habitats Directive are stated in Article 2(1) as being to 'contribute towards ensuring bio-diversity through the conservation of natural habitats and of wild fauna and flora in the European territory of the Member States'.[11] Under Article 2(2), measures taken under the Directive are accordingly designed to maintain or restore natural habitats and species of 'Community interest' at favorable conservation status. In the pursuit of these objectives, the Habitats Directive advances a two-pronged approach to the conservation of European fauna and flora. First, it provides for the creation of a network of Special Areas of Conservation (SACs), known collectively as 'Natura 2000'. The Natura 2000 network consists of sites identified by Member States as comprising particular habitat types (listed in Annex I of the directive), as well as the habitats of particular species (listed in Annex II). Secondly, under Article 2(1) Member States are required to establish a system for the strict protection, within their natural range, of animal species that are listed in Annex IV(a) of the directive and thereby ensure that the distinct conservation and management requirements established for such species are observed throughout their territory.

The most visible aspect of the conservation and management policies pursued under the auspices of the Habitats Directive is the establishment of the Natura 2000 network. The practical mechanics of this project are laid down in Article 4 of the Directive, whereby it is incumbent upon Member States to propose a list of appropriate sites containing the natural habitat types listed in Annex I, as well as those that host species listed in Annex II, for which the Commission will then draw up a draft list of Sites of Community Importance (SCIs). Following this, the Member State in question is required to officially designate any such site within their jurisdiction as a SAC. In 2008 a list of SCIs pertinent to Estonia was published,[12] from which the nature conservation authorities are currently developing national SACs.

The Nature Conservation Act and EU law

The Nature Conservation Act, which was subsequently modified in June 2004, seeks under Article 1(1) to 'protect the natural environment by promoting the preservation of biodiversity through ensuring the natural habitats and the populations of species of wild fauna, flora and fungi at a favourable conservation status'.[13] To this end, nature conservation is to be realized through the establishment of protected areas and 'based on the principles of balanced and sustainable development'.[14]

In discharging these obligations, the Act provides for a dizzying array of different categories (sometimes also sub-categories) of protected areas, which presents what is, in practice, a complex and expansive network of legally recognized sites. By way of summary, the following six broad areas may be established under Article 4(1) by the national authorities in respect of 'protected natural objects': protected areas; special conservation areas; protected species, fossils and minerals; species protection sites; individual protected natural objects; natural objects protected on a local governmental level.

Of the six areas for protected objects envisaged by the Act, the most significant is undoubtedly 'protected areas', which, under Article 4(2), are to be 'maintained in a state unaltered by human activity or used subject to special requirements where the natural environment is preserved, protected, restored, researched or introduced'.[15]

Three types of 'protected areas' are prescribed by the Act. Firstly, under Article 26(1), 'national parks' are protected areas 'prescribed for the preservation, protection, restoration, research and introduction of the natural environment, landscapes, cultural heritage and balanced use of the environment of the protected area'.[16] A non-exhaustive list of national parks is provided, specifying Lahemaa, Karula, Sooma, Vilsandi and Matsalu as national parks recognized under the Act. Within the realm of national parks, such areas may be further delineated into strict nature reserves, special management zones and limited management zones, in which the degree of permitted economic and industrial activity is established with reference to the legal strength of these individual areas.

Secondly, under Article 27(1), 'nature conservation areas' are protected areas 'prescribed for the preservation, protection, restoration, research and introduction of the natural environment', within which the government may also allocate the zones applicable to national parks. The third sub-category of protected area is a 'landscape protection area', defined under Article 28(1) as 'an area prescribed for the preservation, protection, restoration, research, introduction and regulation of use of landscapes of the protected area'. Such areas are predominantly designed for forest habitats, within which special management zones and limited management zones may be allocated, but for which the stronger protection of a strict management zone is precluded. In relation to the other types of 'protected natural objects' established under Article 4, 'special conservation areas' are also significant and by virtue of Article 32(1) are set aside to ensure a favorable conservation status for fauna, flora and fungi.

The designation process and the protected areas envisaged under the Nature Conservation Act – as well as the protection afforded to species – is essentially in keeping with the demands of the Habitats Directive. Indeed, although the range of protected areas established under the Nature Conservation Act regime are rather complex and convoluted – and, indeed, must be read in conjunction with other related statutes addressing forestry concerns, fresh and saltwater areas and agricultural land – it can be seen that there is admirable scope for discharging the SAC obligations under the Habitats Directive, in which economic and industrial activities are not necessarily precluded and, in fact, often widely tolerated (Nollkaemper 1997, p. 271). Moreover, the Nature Conservation Act contains specific provisions for the development of the Natura 2000 network, with Estonia pledging to establish SPAs and SACs in line with Community legislation (Article 69), and to maintain such areas at a favorable conservation status (Article 70).

In this regard, Estonia appears to be in a strong preliminary position to fulfill the designation requirements of the two directives, and the Nature Conservation Act may be considered a useful legislative framework by which to attain this aim. By way of clear evidence, since accession Estonia has designated 66 separate SPAs under the Wild Birds Directive, comprising a total area of 12,160.8 km^2 or 12.8% of national territory, while 509 SCIs were declared under the Habitats Directive, comprising 10,591.1 km or 15.9% of national territory, the highest ratio of protected areas to total land mass of all the 2004 entrants to the EU.[17] In total, the Estonian authorities have designated some 683,099 hectares of territory as protected areas (Klein 2008, p. 28), ranging from the vast Lahemaa National Park to the miniscule 'Roheline Turg' in Tallinn Old Town, which is among the tiniest protected areas in Europe. Nevertheless, the designation process is only an initial stage in the life of a protected area, with strong continuing commitments towards the maintenance and surveillance of protected habitats and species prescribed under the Habitats Directive. With the Natura 2000 network starting to expand throughout the Community, a number of Member States have been censured by the Commission and the European Court of Justice (ECJ) for failing to adhere to the strict monitoring and protection requirements under these directives.

Accordingly, the position may explain many of the concerns raised by the Commission in relation to the LPNO during the candidacy process which, as noted above, essentially created a series of 'paper parks' endowed with modest resources and starved of the necessary additional administrative measures in order to function correctly. That the Soviet-style system of further secondary legislation has been swept away is therefore a considerable practical improvement. However, additional concerns may be expressed in relation to the current operation and administration of the Nature Conservation Act, which will need to be addressed if Estonia is to avoid central EU compliance mechanisms. In this respect, three key problems have been highlighted by the most recent and comprehensive review of Estonian biodiversity practices, made in conjunction with international commitments under the Convention on Biological Diversity (Riigikogu 2008, p. 7).

First, as highlighted throughout the accession process, the enforcement of the provisions under the Nature Conservation Act is generally weak and the sanctions prescribed offer a limited degree of deterrence. Indeed, by the government's own admission, '[t]he current intensive expansion of protected areas in the conditions of weak and soft administrative management is not effective enough. It is important that legislation regulating the use of the natural environment is obeyed without question' (Riigikogu 2005, p. 31). Second, nature conservation remains poorly resourced and the protection and sustainable use of biodiversity 'is not always a priority to those in charge', while funding is often insufficient to meet current needs (Riigikogu 2008, p. 8). Allied to this, administrative difficulties – which have been a consistent feature of the Soviet and post-Soviet eras – have yet to be fully rectified, with a high turnover of staff in relevant government agencies (Riigikogu 2008, p. 8). Both problems clearly have a negative impact upon the ability of the pertinent bodies to effectively monitor and protect such areas in practice. Third, management plans have not been widely concluded for the majority of key protected areas (Riigikogu 2008, p. 44), hence a coordinated system of conservation targets for many such species and habitats remains rather conspicuous by its absence.

Such shortcomings may prove significant if viewed in the light of recent European Court of Justice (ECJ) practice, in which a complaint containing the 'perfect storm' of the three key deficiencies outlined above was raised by the Commission against Ireland.[18] In this case, a series of alleged violations of the Habitats Directive in relation to an eclectic group of Irish species was upheld, with particular criticism reserved for the failure to elaborate species action plans, considered 'an effective means of meeting the strict protection requirement'[19] laid down in the directive, as well as the 'especially meagre'[20] resources allocated for monitoring and surveillance activities deemed particularly grave. If a similar complaint were to arise in relation to a number of Estonian protected areas at present, a similarly strident outcome might not be considered overly surprising.

International nature conservation commitments

In addition to the keystone nature conservation directives adopted by the EU, a series of binding external commitments are also incumbent upon Estonia by virtue of participation in a host of international environmental agreements, many of which address nature conservation concerns. Given their vast number and broad range, it is not possible to advance a critical overview of national practice within the full range of such bilateral and multilateral instruments to which Estonia is a party. Instead, this section will examine the impact of membership of a numerically small but highly significant group of treaties that have acquired global prominence in the field of nature conservation. In this respect, it will be noted that participation has created free-standing obligations, which are essentially transposed through the current corpus of nature conservation legislation, while EU practice in relation to such treaties has also had a strong impact upon their application to Estonia.

Prior to examining each treaty in turn, it should be observed that the dissolution of the USSR had particular implications for the Baltic states with regard to an expansive range of conventions which, by virtue of the 1940 annexation, entailed *de facto* commitments for the Estonian SSR. The dissolution of states presents a series of difficulties as far as succession to treaty obligations is concerned (Crawford 2006, pp. 375–449). In the case of the USSR, despite some initial confusion (Aust 2000, p. 314), Russia is considered the successor state and accordingly assumes the international commitments to which the Soviet Union was previously bound (Blum 1992, p. 354). The precise scope of the obligations assumed by the modern Republic of Estonia in relation to the first Republic is an issue that has raised significant practical difficulties (Mälksoo 2003), although international environmental obligations from the inter-war period are now rather obsolete and nature conservation issues are now relatively non-controversial (Williams 1994, p. 1).

In addition to the legal dilemmas posed in relation to successor states, considerable practical problems were also raised by treaty-making in the immediate post-Soviet years. A number of conventions to which the USSR had originally been party were also attractive propositions for the newly independent authorities. Nevertheless, the practical mechanics of treaty accession in the early years of the new regime presented substantial challenges for those charged with facilitating membership and implementation of these instruments, creating a significant administrative bottleneck. However, these initial teething troubles aside, Estonia has subsequently become a party to a considerable array of international instruments, for which the following are likely to prove most influential in framing present and future nature conservation policies.

The Convention on Biological Diversity

The most logical – if not chronological – starting point for a discussion of pertinent international commitments is the Convention on Biological Diversity (CBD). The CBD was developed under the auspices of the UN Conference on Environment and Development in Rio de Janeiro in 1992 and remains one of the most widely ratified multilateral instruments currently addressing nature conservation concerns, with 191 parties to date. Estonia signed the Convention in June 1992 and became a formal party on 27 July 1994.

The CBD is a lengthy document, but aims primarily ‘to anticipate, prevent and attack the causes of significant reduction or loss of biological diversity at source’ (preamble), by promoting cooperation, institution-building and the conservation of ecosystems and natural habitats and the recovery of endangered species. To this end, while recognizing the sovereign right of states to exploit their own resources pursuant to national environmental policies, such rights are intertwined with the responsibility to ensure that activities conducted within their jurisdiction do not cause damage to other states (Article 3). Under Article 6, parties are primarily required to develop national strategies, plans and policies for the conservation and sustainable use of biodiversity and to integrate such strategies into relevant sectoral policies. To attain such goals, parties are required to cooperate through competent international organizations (Article 5), as well as to establish protected areas (Article 8) and conduct research programs (Article 9). As noted above, a number of these key principles are echoed in the Estonian constitution, while the Nature Conservation Act 2004 provides an operational framework by which to establish a network of protected areas in a manner that will address commitments under both EU law and the CBD.

Taken on its own terms, the CBD is not an instrument that prescribes clear and unequivocal obligations upon its constituent parties. Instead, it acts primarily as a framework regime to foster a national culture towards a wide implementation of sustainable practices across all sectors of activities. Insofar as specific objectives are prescribed, the most important commitment under the CBD in recent years was established at the Sixth Conference of the Parties in 2002, whereby it was declared that '[p]arties commit themselves to a more effective and coherent implementation of the three objectives of the Convention, to achieve by 2010 a significant reduction of the current rate of biodiversity loss at the global, regional and national level as a contribution to poverty alleviation and to the benefit of all life on earth' (Decision VI/26). As noted above, this provision has had a strong galvanizing effect on EU biodiversity policies. Accordingly, in addition to national strategies, Estonia will be required to adhere to Community-wide programs to address this target, given that the EU is itself a party to the CBD and strives to reduce biodiversity losses throughout the Member States as a collective whole.

To date, the Estonian implementation of CBD commitments appears to have been decidedly mixed, hence the influence of the current EU programs as a regulatory spur may prove to be a welcome external pressure from the point of view of nature conservation. As a practical means of fulfilling national obligations under Article 6, parties to the CBD are required to develop National Biodiversity Strategies and Action Plans (BSAPs); the Estonian BSAP was adopted in 1999 and is currently undergoing revision in the light of further guidance elaborated by the CBD at the Ninth Conference of the Parties (Decision IX/8). Moreover, under Article 26 parties are required to report periodically upon the implementation of national BSAPs, with the next reporting deadline set at March 2009. Estonia tendered its report in December 2008, becoming in fact the first CBD party to do so. While the report revealed that a substantial body of work had been undertaken in establishing protected areas (Riigikogu 2008, p. 7), significant shortcomings were revealed – not least in the training on sustainable development issues provided to government employees and the reticence accorded by the *Riigikogu* to the first BSAP, operational between 1999 and 2005, which essentially viewed the document as a voluntary set of standards and an implementation program that was 'ahead of its time' (Riigikogu 2008, pp. 47–8).

The Convention on International Trade in Endangered Species of Flora and Fauna

A further instrument of considerable significance is the Convention on International Trade in Endangered Species of Flora and Fauna (CITES), which was adopted in Washington in March 1973 and entered into force on 1 July 1975. The CITES regime is one of the most widely supported biodiversity-related conventions and currently has 175 parties. Estonia signed the convention in July 1992 and became a formal party on 20 October 1992.

CITES aims to provide a forum for international cooperation to protect certain species of flora and fauna from over-exploitation through trade. To this end, species are listed within three broad Appendices to the convention, which provide varying degrees of protection and trade restrictions. Under Article II(1), species listed in Appendix I are those threatened with extinction or that may be affected by trade, therefore economic activities must only be authorized under exceptional circumstances. Under Article II(2), Appendix II species are not necessarily threatened with extinction, but may become so unless subject to trade restrictions. Species listed in Appendix III have a more subjective designation and are listed as subject to regulation within an individual jurisdiction, with international cooperation necessary to ensure that their conservation status is not undermined by trade [Article II(3)]. In Estonia, trade issues are addressed under two key provisions of the Nature Conservation Act 2004. First, under Article 56(2) where a species is listed as protected under national law, any sale of such a specimen, or offences inchoate thereto, is prohibited. Likewise, Article 59 prevents the transnational trade in such species, specifically citing the EU CITES Regulation[21] as the basis for these provisions.

Two key issues may be identified in relation to Estonian practice within CITES. In the first instance, a review of the enforcement capabilities of the post-2004 EU Member States revealed that Estonia, while enjoying a good level of cooperation with neighboring countries in addressing the trade in endangered species, nevertheless had no national enforcement group, with monitoring activities undertaken between the police, customs and veterinary officials with little formal coordination (Parry-Jones & Knapp 2005, p. 35). This is significant for Community, as well as national, policies on wildlife crime, given that the adjustment of the Schengen boundaries means that Estonia now guards the eastern borders of the EU. Accordingly, some further consolidation of the national enforcement system may be required in order to ensure that the current EU-wide protection regime is not undermined, given that the effect of the common market essentially permits goods to be freely traded within the Community.

Second, Estonian policies within CITES will be increasingly driven by the European Commission in future years. At present, unlike the CBD, the EU is effectively precluded from formal membership of CITES. Indeed, at the time of the conclusion of the Convention in 1973, given that the environmental competence of the EEC could be best considered nascent (Davies 2004, pp. 1–25), no provision was made within CITES for participation by Regional Economic Integration Organizations in a manner that is now relatively routine in environmental accords. In 1983 an extraordinary meeting of CITES resulted in the amendment of Article XXI, which now permits Regional Economic Integration Organizations to accede, under what is popularly referred to as the 'Gabarone Amendment' to the Convention. Nevertheless, the EU currently remains outside the official umbrella of CITES, as the amendment has to date failed to meet the requirements for entry into force as prescribed under Article XVII(3), namely the acquiescence of two-thirds of the states that were parties to the Convention at the material time. Instead, the Commission in practice establishes an aggregated position of its constituent Member States prior to any meeting of the CITES institutions, through which one EU party will advance the Community agenda by introducing motions as if it were acting in its own interest. While Member States are able to advance policies within CITES that also correspond to the more domestic concerns, it appears that Estonian participation within the Convention structure in future years will be primarily as an additional supporting voice to a wider Community agenda.

The Convention on the Conservation of Migratory Species of Wild Animals

Given the strategic importance of Estonia as a key area of habitat and pathway for migratory birds, the Convention on the Conservation of Migratory Species of Wild Animals (hereinafter the 'CMS' or 'Bonn Convention') might be considered a treaty regime of particular significance. An international instrument for the conservation of migratory species of wild fauna was first mandated at the Stockholm Conference on the Human Environment in 1972 and, following an initial struggle to accumulate participants, has become a well-supported treaty in recent years with 110 parties to date, of which Estonia formally became a member on 1 October 2008. The CMS entered into force on 1 November 1983 following ratification by the requisite 15 parties (Caddell 2005, p. 113).

The CMS recognizes that successful conservation and management measures in relation to migratory animals requires 'the concerted action of all States within the national jurisdictional boundaries of which such species spend any part of their life cycle' (preamble). To this end, the CMS adopts a two-tier approach, with the parties drawing a distinction between species identified as 'endangered' (listed in Appendix I) and those considered to have an 'unfavourable conservation status' (listed in Appendix II), with differing obligations and policies prescribed in relation to each category. In so doing, strict protection measures are prescribed for such species, including regulating anthropogenic activities that may cause harm to such species and/ or their habitats (Article III).

The CMS is perhaps most notable for its innovative regulatory system in respect of Annex II species, which encourages the parties to cooperate to advance the conservation of such species, by means of subsidiary agreements adopted under the auspices of Article IV of the Convention. In this respect two broad types of instruments are envisaged: AGREEMENTS under Article IV(3) and Agreements under Article IV(4) which prescribe differing obligations and criteria for participations.[22] Due to this arrangement, a participant to a CMS subsidiary Agreement need not be a formal party to the parent convention, as such instruments are essentially designed to be self-sufficient offshoots from the Bonn Convention. This may partly explain the comparative reticence of Estonia to commit to full membership of the CMS, given its long-standing participation in the African-Eurasian Waterbird Agreement (AEWA) and the Agreement on the Conservation of Populations of European Bats (EUROBATS), which remain the migratory species most pertinent to Estonian ecosystems (Caddell 2008, pp. 254–9). Estonia has additionally signed the Agreement on the Conservation of Small Cetaceans of the Baltic, North-East Atlantic, Irish and North Seas (ASCOBANS). In 2006 it informed both the Advisory Committee and the Meeting of the Parties that formal ratification was expected imminently, but in reality little tangible progress had been made by the time of the Fifteenth Meeting of the ASCOBANS Advisory Committee in 2008.

In future years, Estonian implementation of its CMS commitments, especially those relating to AEWA and EUROBATS (as well as any potential obligations raised by eventual ratification of ASCOBANS), is likely to be directed through national action under the Habitats Directive. Indeed, the EU maintains a somewhat cool relationship with the various CMS institutions and its subsidiaries (Caddell 2008, p. 256). As far as bats and cetaceans are concerned, 'all species' of these mammals are listed in Annex IV(a) of the Habitats Directive, for which Estonia is obliged to maintain a system of strict protection. Likewise, in the context of AEWA, national activity will be directed through the Wild Birds Directive. The EU is a formal party to this Agreement (but not to ASCOBANS or EUROBATS), but only insofar as any commitments prescribed under AEWA are 'consistent with, and do not entail modification of, Community legislation on the conservation of wild birds and their natural habitats'.[23] Moreover, the aggregation of a Community position within such instruments – including those within which the EU has no intention of participating – is likewise a strong feature of Member State practice, with an EU coordination meeting held prior to each day of negotiations, much to the chagrin of non-EU Member States and observers (Caddell 2008, pp. 256–7).

The International Convention for the Regulation of Whaling

One very recent development in the international context has been the accession of Estonia to the International Convention for the Regulation of Whaling (ICRW) on 7 January 2009. The ICRW is a highly controversial treaty, adopted in 1946, with the aim of advancing through its constituent regulatory forum, the International Whaling Commission (IWC), 'the proper conservation of whale stocks and thus [making] possible the orderly development of the whaling industry' (preamble). Estonia has no national interest in the hunting of whales, although the IWC has in recent years examined aspects of the conservation of the Baltic harbour porpoise, a species known to fleetingly inhabit national waters, and one that is protected under Annex II of the Habitats Directive.

Given these relatively superficial interests in whales, the significance of Estonian accession to the Whaling Convention is strongly illustrative of the external policies of the European Commission and the further aggregation of a Community position within key international fora. As with CITES, the current terms of the ICRW preclude accession to the treaty by a Regional Economic Integration Organization, hence formal membership of the IWC would require an amendment of the Convention. Given that the European Council has repeatedly blocked moves towards EU accession to the treaty (Churchill 2001, p. 234), the Commission in 2008 tabled a Communication to the Member States seeking to harmonize a common stance within the IWC on certain issues such as the desirability of an ongoing global moratorium on commercial whaling, and the need to discourage so-called 'scientific whaling' conducted by various fleets under a loophole within the Convention. Accordingly, Estonian accession may be viewed as a further example of adherence to the growing EU aggregation of a common negotiating position within major multilateral biodiversity regimes.

The Convention on the Protection of the Marine Environment of the Baltic Sea Area

On a regional level, an agreement of substantial significance to Estonian nature conservation efforts – with particular reference to the marine sphere – is the Convention on the Protection of the Marine Environment of the Baltic Sea Area 1992, which entered into force on 17 January 2000. The Helsinki Convention is one of the UNEP Regional Seas Agreements established to provide a regulatory framework and a basis for regional cooperation in respect of key sea areas around the globe. The first Helsinki Convention was adopted in 1974 and was subsequently renegotiated following the dissolution of the Soviet Union and the re-emergence of the Baltic states.

The fundamental principles of the Helsinki Convention are established in Article 3, which requires the parties to jointly or individually take appropriate legislative, administrative or other relevant measures to prevent and eliminate pollution 'in order to promote the ecological restoration of the Baltic Sea Area and the preservation of its ecological balance'. In so doing, the parties are to take a precautionary approach to addressing pollution concerns from a host of vessel- and land-based sources (Articles 4 to 15). A designated body, the Baltic Marine Environment Protection Commission (HELCOM), is established to oversee and direct such efforts.

The significance of the HELCOM regime to Estonia should be viewed in the light of the recent adoption by the EU of the Marine Strategy Framework Directive.[24] Under Article 1(1) the MSFD seeks to provide 'a framework within which Member States shall take the necessary measures to achieve or maintain good environmental status within the marine environment by the year 2020 at the latest'. To this end, a 'good environmental status' involves the provision of 'ecologically diverse and dynamic oceans and seas which are clean, healthy and productive within their intrinsic conditions, and the use of the marine environment is at a level that is sustainable, this safeguarding the potential for uses and activities by current and future generations' [Article 3(5)]. Implicit in this optimal condition is that constituent marine ecosystems can withstand anthropogenic change and habitats and species are protected, while the anthropogenic impact of substances and energy in the marine environment does not cause pollution effects. In ascertaining the environmental status of Community seas, a series of indicators and qualitative descriptors are established in the Annexes to the directive.

In pursuing this objective, the MSFD largely eschews a 'top-down' Community-level approach and, in keeping with the principle of subsidiarity, places responsibility for marine governance primarily at a national or regional level, subject to EU supervision. This approach, in essence, constitutes an important recognition that generic responses to marine environmental problems are of limited value in addressing the four broad areas of Community seas, all of which face different ecological pressures. Accordingly, in the first instance, under Article 5(1) each Member State is required to develop a marine strategy for its national waters, encompassing a clear assessment of their current environmental status and a targeted program of measures to be introduced by 2016 at the latest. Recognizing that individual coastal states are components of a wider marine region or sub-region, Member States are to take 'due account' of this position [Article 4(1)] and cooperate to ensure that a good environmental status is attained in respect of the region or sub-region concerned [Article 5(2)]. In implementing these commitments, under Article 8(1) national assessments should examine essential features and characteristics of these areas, the predominant pressures upon them and their primary economic and social uses. From this appraisal a series of environmental targets shall be identified (Article 10), as well as coordinated monitoring programs for the ongoing assessment of these waters. Article 13 of the MSFD requires Member States to develop a detailed program of measures. As part of this package, the objectives of these programs should make some contribution to further developing the protected areas established under the Wild Birds Directive and the Habitats Directive, rendering the MSFD co-existent with and complementary to pre-existing commitments established under EU nature conservation law and policy.

A particular strength of the directive is its emphasis upon the use of existing regional structures, focusing upon the potential application of the Regional Seas Agreements, with the EU keen to avoid the duplication of commitments that largely blights the operation of key multilateral environmental agreements in practice. In this respect, it appears that Estonian policy in relation to the marine environment in future years will be centered on the further development of regional standards in relation to the Baltic Sea with the overall aim of replenishing this degraded marine zone and restoring it to a 'good environmental status'.

Concluding remarks

Estonia has experienced an unheralded yet quietly significant history as a nature conservation actor throughout the twentieth century, both during the inter-war period and also as an individual component of the USSR. In this respect, the Estonian SSR, although often stymied by a rigid Soviet structure and with limited opportunity to advance norms on its own terms, nonetheless remained a strong advocate of further conservation measures throughout the USSR. That such calls were largely unheeded is testament to the often rather illusory protection granted to *zapovedniki* during the Soviet era – not least under Stalin's administration – as well as to the wholesale inefficiency brought by the excessively centralized policies that characterized Soviet economic and development planning and environmental protection norms.

Since the fall of Communism, Estonia has been keen to advance nature conservation laws, not only to consolidate existing positive structures upon independence, but also to address deficiencies in coverage during the Soviet era. Latterly, as potential EU candidacy became a strong motivating factor, biodiversity governance became centered on the harmonization of existing national norms with the Habitats Directive, both as a precursor to EU entry and the mandatory acceptance of the environmental *acquis*, as well as following an admirable model as a legislative guide.

Likewise, post-independence, Estonia has become party to a number of key conservation treaties. Participation in such regimes has had a generally positive effect, both in terms of re-engaging with the international community following Soviet annexation, as well as shaping further areas of conservation activity. In the coming years, the multilateral regime of most immediate importance to Estonia seems likely to be HELCOM, as it embarks upon an ambitious program of activity to address the highly degraded state of the Baltic Sea, in which it is strongly supported by the regulatory mandate provided by the EU under the Marine Strategy Framework Directive.

In future years, EU membership is likely to be the guiding theme in Estonia's external biodiversity commitments. The Commission will undertake rigorous monitoring not only of national compliance with the Habitats Directive and Wild Birds Directive, for which Estonia has already designated an impressive network of protected areas, but also in relation to international commitments, for which the implementation and policy direction are being increasingly driven by a strong EU presence. In the biodiversity field, perhaps more than in any other multilateral environmental sector to date, the Commission has sought to aggregate a common stance on the negotiation of targets and the ultimate direction of such treaties. Indicative of this position, perhaps, is the recent Estonian accession to the Whaling Convention, while the national position within CITES and the CMS is also likely to be increasingly Commission-driven.

Despite this increasing international and supranational focus, nature conservation essentially remains a task for national actors, for which the coming decade is likely to prove challenging. In many respects, the designation of protected areas is a preliminary aspect of Habitats Directive practice; the directive prescribes strict ongoing obligations for the monitoring and protection of such sites and species. Indeed, the quotidian administration of Natura 2000 sites entails a substantial commitment of both financial and practical measures, with which Estonia will need to invest wisely if the current admirable program of networks is to remain more than a series of 'paper parks'. Accordingly, given the practical and legal challenges that lie ahead, while the history of nature conservation legislation in Estonia may be traced back nearly 800 years, in a number of key respects the story of national biodiversity protection may be only just beginning.

Notes

1 The Natura 2000 network is a network of Special Areas of Conservation designated by the EU Member States under the Habitats Directive that creates an EU-wide program of protected areas of Community importance. All Member States are expected to contribute to the Natura project, which to a considerable extent explains the strong emphasis on the reform of national nature conservation legislation in fulfilling the environmental *acquis* as a key part of the accession process.

2 Council Directive 92/43/EEC of 21 May 1992 on the conservation of natural habitats and of wild fauna and flora [1992] *Official Journal* L206/7. The most recent consolidated version of the Directive was adopted on 1 January 2007, following the latest round of accession to the EU, incorporating Bulgaria and Romania into a further enlarged EU of 27 Member States.

3 See http://www.eel.nl, accessed 1 June 2009.

4 Article 6 states, '[e]nvironmental protection requirements must be integrated into the definition and implementation of the Community policies and activities referred to in Article 3, in particular with a view to promoting sustainable development'.

5 The *acquis* denotes a substantial body of existing EU legislation, policy and case-law for which the legislation of the candidate country must be essentially similar by the time at which it accedes as a Member State.

6 Council Directive 79/409/EEC of 2 April 1979 on the conservation of wild birds [1979] *Official Journal* L103/1.

7 Council Directive 92/43/EEC of 21 May 1992 on the conservation of natural habitats and of wild fauna and flora [1992] *Official Journal* L206/7. The most recent consolidated version of the Directive was adopted on 1 January 2007,

following the latest round of accession to the EU, incorporating Bulgaria and Romania into a further enlarged EU of 27 Member States.

8 COM (1998) 42.

9 Decision No 1600/2002/EC [2002] *Official Journal* L242/1.

10 See note 6.

11 See note 7.

12 Council Decision 2008/24/EC of 12 November 2007 adopting, pursuant to Council Directive 92/43/EEC, a first updated list of sites of Community importance for the Boreal biographical region [2008] *Official Journal* L12/118.

13 Article 1(1) of the Nature Conservation Act.

14 Article 2 of the Nature Conservation Act.

15 Article 4 of the Nature Conservation Act.

16 Article 26(1) of the Nature Conservation Act.

17 European Commission, MEMO/06/292 of 18 July 2006, p. 1.

18 *Commission* v. *Ireland*; Case C-183/05.

19 Opinion of Advocate-General Léger, at paragraph 39.

20 Opinion of Advocate-General Léger, at paragraph 69.

21 Council Regulation 338/97/EC of 3 March 1997 on the protection of species of wild fauna and flora by regulating trade therein [1997] *Official Journal* L61/1.

22 The term AGREEMENT is capitalised due to a quirk of the Convention, which provides for the adoption of two differing forms of regional instrument, both termed 'agreements' but which operate under slightly different conditions.

23 Article 3 of Council Decision 2006/871/EC;[2006] *Official Journal* L345/25.

24 Directive 2008/56/EC of the European Parliament and of the Council of 17 June 2008 establishing a framework for community action in the field of marine environmental policy [2008] *Official Journal* L164/19 [Hereinafter 'MSFD'].

References

Andresson, A. & Kolk, T. (2003) 'The Role of Basic Rights in Environmental Protection: Basic Right to Environment *de lege feranda* in the Estonian Constitution', *Juridica International*, 8, 1, p. 140.

ASCOBANS (2004) *Report of the Eleventh Meeting of the Advisory Committee to ASCOBANS* (Bonn, UNEP).

Auer, M. R. (2005) 'Estonian Environmental Reforms: A Small Nation's Outsized Accomplishments', in Auer, M. R. (ed.) (2005) *Restoring Cursed Earth: Appraising Environmental Policy Reforms in Eastern Europe and Russia* (Oxford, Rowman & Littlefield).

Auer, M. R. & Raukas, A. (2002) 'Determinants of Environmental Clean-Up in Estonia', *Environment and Planning C: Government and Policy*, 20, 5, p. 679.

Aust, A. (2000) *Modern Treaty Law and Practice* (Cambridge, Cambridge University Press).

Blum, Y. Z. (1992) 'Russia Takes over the Soviet Union's Seat at the United Nations', *European Journal of International Law*, 3, 2, p. 354.

Caddell, R. (2005) 'International Law and the Protection of Migratory Wildlife: An Appraisal of Twenty-Five Years of the Bonn Convention', *Colorado Journal of International Environmental Law and Policy*, 16, 1, p. 113.

Caddell, R. (2008) 'Biodiversity Loss and the Prospects for International Cooperation: EU Law and the Conservation of Migratory Species of Wild Animals', *Yearbook of European Environmental Law*, 8, p. 218.

Churchill, R. (2001) 'Sustaining Small Cetaceans: A Preliminary Evaluation of the Ascobans and Accobams Agreements', in Boyle, A. & Freestone, D. (eds) (2001) *International Law and Sustainable Development: Past Achievements and Future Challenges* (Oxford, Oxford University Press).

Crawford, J. (2006) *The Creation of States in International Law* (Oxford, Oxford University Press).

Davies, P. G. G. (2004) *European Union Environmental Law: An Introduction to Key Selected Issues* (Aldershot, Ashgate Publishing).

de Sadeleer, N. (2005) 'Habitats Conservation in EC Law—From Nature Sanctuaries to Ecological Networks', *Yearbook of European Environmental Law*, 5, p. 215.

de Sadeleer, N. (2007) 'The Birds, Habitats and Environmental Liability Directives to the Rescue of Wildlife under Threat', *Yearbook of European Environmental Law*, 7, p. 36.

Dhondt, N. (2003) *Integration of Environmental Protection into Other EC Policies: Legal Theory and Practice* (Groningen, Europa Law Publishing).

Donald, P., Sanderson, F. J., Burfield, I. J., Bierman, S. M., Gregory, R. D. & Waliczky, Z. (2007) 'International Conservation Policy Delivers Benefits for Birds in Europe', *Science*, 317, 5839, p. 810.

European Commission (1997) *Agenda 2000—Commission Opinion on Estonia's Application for Membership of the European Union*; Doc/97/12 of 15 July 1997.

European Commission (2001) *2001 Regular Report on Estonia's Progress Towards Accession*; SEC (2001) 1747.

European Commission (2002) *Towards the Enlarged Union—Strategy Paper and Report of the European Commission on the progress towards accession by each of the candidate countries*; SEC (2002) 1403.

European Commission (2003) *Comprehensive Monitoring Report on Estonia's Preparations for Membership*; COM (2003) 675 final.

Hardin, G. (1968) 'The Tragedy of the Commons', *Science*, 162, 3859, p. 1243.

Kalda, A., Kannukene, L. & Leis, M. (1992) 'Rare Mosses in Estonia and their Protection', *Biological Conservation*, 59, 2–3, p. 201.

Kallas, A. (2002) 'Public Forestry Policy Making in Post-Communist Estonia', *Forest Policy and Economics*, 4, 4, p. 323.

Klein, L. (2008) *Diversity of Nature in Estonia: Estonian Nature Conservation in 2007* (Tallinn, Estonian Environmental Information Centre).

Krämer, L. (2002) 'Thirty Years of European Environmental Law: Perspectives and Prospectives', *Yearbook of European Environmental Law*, 2, p. 155.

Kumari, E. (1971) 'Two Conferences on Environmental Conservation in Estonia', *Biological Conservation*, 3, 4, p. 317.

Kumari, E. (1972) 'Nature Conservation, a Vital Need in Estonia', *Biological Conservation*, 4, 3, p. 230.

Mälksoo, L. (2003) *Illegal Annexation and State Continuity: The Case of the Incorporation of the Baltic States by the USSR* (Dordrecht, Martinus Nijhoff).

Misiunas, R. & Taagepera, R. (2006) *The Baltic States: Years of Dependence 1940–1990* (London, C. Hurst & Co).

Nollkaemper, A. (1997) 'Habitat Protection in European Community Law: Evolving Conceptions of a Balance of Interests', *Journal of Environmental Law*, 9, 2, p. 271.

Parry-Jones, R. & Knapp, A. (eds) (2005) *Country profiles compiled for the EU Wildlife Trade Enforcement Co-ordination Workshop, 25–27 October 2005* (London, DEFRA).

Pryde, P. (1972) *Conservation in the Soviet Union* (Cambridge, Cambridge University Press).

Pryde, P. R. (1986) 'Strategies and Problems of Wildlife Preservation in the USSR', *Biological Conservation*, 36, 4, p. 351.

Riigikogu (2005) *Sustainable Estonia 21* (Tallinn, Riigkogu).

Riigikogu (2008) *IV National Report to the Convention on Biological Diversity: Estonia* (Tallinn, Riigkogu).

Saarniit, L. (2004) *Multilevel Governance: The Implementation of the Habitats Directive in Estonia* (Dublin, Organisation for EU Enlargement Occasional Papers).

Sepp, K., Palang, H., Mander, U. & Kaasik, A. (1999) 'Prospects for Nature and Landscape Protection in Estonia', *Landscape and Urban Planning*, 46, p. 161.

Taagepera, M. (1989) 'The Ecological and Political Problems of Phosphate Mining in Estonia', *Journal of Baltic Studies*, 20, 2, p. 165.

Veinla, H. (1996) 'Environmental Considerations in Estonian Land-Use Legislation', *Juridica International*, 1, 1, p. 48.

Williams, P. (1994) 'The Treaty Obligations of the Successor States of the Former Soviet Union: Do They Continue in Force?', *Denver Journal of International Law and Policy*, 23, p. 1.

GREEN, BLACK AND BROWN: UNCOVERING LATVIA'S ENVIRONMENTAL POLITICS

David J. Galbreath and Daunis Auers

Introduction

The Latvian Green Party (*Zaļa Partija*, ZP), as part of the Green/Farmers' Union (*Zaļo un Zemnieku Savienība*, ZZS), has been an important actor in Latvian politics. As part of a minority government led by ZZS, Indulis Emsis was the world's first Green party prime minister, although his government only lasted for ten months. Indeed, ZZS has been in every government coalition since November 2002. Despite this success in Latvian party politics, environmentalism has become increasingly depoliticized since the pro-environmental movement in the late Soviet period set the precedent for a nationalist movement that would challenge the Soviet regime. For example, the Latvian Green Party's past electoral rhetoric illustrates a clear lack of 'Green' initiatives. Thus, the situation in Latvia illustrates a conundrum that leaves us with several questions concerning the Baltic context in general and Latvia specifically. Despite active environmental movements in the late Soviet period, why have we seen a depoliticization of environmentalism? What is the nature of environmental attitudes in Latvia now? How has the Latvian Green Party managed to remain an important element in post-Soviet society, even to the point of controlling government? The answer to this question can be sought in the many sides to ZZS: whether environmental, nationalist or corporate.

We begin with the argument that there are three sides to the environmental movement in Latvia. These sides are illustrated by their respective colors: green, brown and black. By 'green', we mean the ecological movement for the preservation of the natural environmental and reversal of harmful side-effects of industrialization. By 'brown', we mean civic and ethnic nationalisms. Finally, by 'black', we mean the role of the oil transit industry in capturing the 'Green' agenda in Latvia. In order to fully appreciate how these different sides relate to one another, we must first conceptualize the relationship between environmentalism, nationalism and special interests in general. From this, we will be able to see how environmentalism in the late Soviet period could result in a mixture of interests represented within one 'Green' label in contemporary Latvia.

Environmentalism and the 'Nation'

First, what do we mean by environmentalism? The environmental movement in North America and Western Europe resulted from the protest movements of the 1960s and 1970s. There was a call not only to halt harm to the natural environment but also to reverse any previous harm. Katrina Z. S. Schwartz (1999, 2006) examines how Eastern notions of nation have led to a strong emphasis on *heimat*, or caretaker societies, as opposed to Western notions of pristine nature. As we shall see in the next section, Latvian environmental discourses vary between the reformist political and economic agendas and the radical social agendas of nationalism. For this reason, Latvia is an interesting case of overlap between environmentalism and nationalism. How do these otherwise opposing political persuasions come to overlap in general and Latvia specifically?

Environmentalism and nationalism are odd bedfellows for two reasons. First, environmentalism in Western Europe and North America has often been identified with the political left. For instance, Green parties have traditionally come out of movements for social justice and have typically supported, or been a part of, social democratic governing coalitions, such as the German 'Red–Green' alliance from 1998 to 2005. Nationalism has traditionally been a movement of the political right. Nationalist parties have primarily supported, or been a part of, conservative governing coalitions, such as Estonia's *Isamaaliit* (*Pro Patria* Union). Secondly, analytically nationalist and environmental movements are seen as 'old' and 'new' social movements, respectively. By 'old' social movements, we mean those with broad popular support and a broad issue base, such as the labor movement. Alternatively, 'new' social movements have been characterized as being single-issue with a substantial, but narrow support base (see Klandermans 1991; Kriesi *et al.* 1992; Larana *et al.* 1998). Thus, it may come as a surprise that in general nationalist movements have borrowed from the environmentalist agenda rather than vice versa. How does this work?

Paul Hamilton (2002) finds that nationalist movements 'mine' environmental discourse to legitimate their political grievances and boost their political support. This harvesting by an old social movement of a new social movement is especially complex in the case of nationalism and environmentalism. Effectively, nationalism works like an ideological parasite and inasmuch as we can discuss environmentalism as an ideology (otherwise referred to as ecologism), we can see a discursive transfer between one and the other. In his analysis of this relationship, Hamilton uses the traditional dichotomy of nationalism: civic and ethnic. By civic nationalism, we mean the prioritization of social identities pertaining to the political entity that is the state. As Hamilton states, civic nationalism allows exit and entry, for one can renounce one's citizenship or acquire another (2002, p. 30). Ethnic nationalism, on the other hand, implies the prioritization of social identities pertaining to an ethnic group. Ethnic nationalism has no such entry or exit, other than birth and death. What binds civic and ethnic nationalism is the ideologies' embedded stake in the territorial state or nation. Thus, there is a natural link between the nation as a people and political entity, and the nation as a land. Herein lies the connection between nationalism and environmentalism.

These two movements potentially converge in their connection to the inhabitation and use of territory. Hamilton (2002, pp. 38–40) provides several ways in which environmentalism and nationalism may overlap. First, local nationalism often seeks democratic and decentralized forms of governance. Such an overlap could be seen in the late Soviet period in the Baltic states. Second, both movements deplore the 'soulless technocratic state' and prioritize smaller political communities, even to the extent of elevating the value of the local village. 'Smaller [communities] might be expected to be better stewards of the land when their states are so small that the effects of environmental irresponsibility are so readily and quickly apparent' (2002, p. 39). The third potential overlap is what Hamilton refers to as 'bioregionalism' or the long-term vision of developing, protecting and passing on the land. This process is most clearly what we see as *heimat*, as discussed earlier.

However, nationalism and environmentalism are in many ways mutually incompatible. Hamilton offers two reasons. First, the local and the global in environmentalism are seen as equally important. In essence, biospheres are connected and in many ways mutually dependent. Nationalism, on the other hand, prioritizes the local over the global. Hamilton states, 'the narcissism of nationalism is necessarily anthropocentric and utilitarian as regards the natural world. This extreme form of nationalism may treasure a place (as the Serbs treasure Kosovo) but see the rest of the world as potential resources' (2002, p. 38). Second, where environmentalism concentrates on spiritual, non-material benefits from the land, nationalism often sees the land as something to be used. This divergence again brings us back to Schwartz's analysis of conservation versus biodiversity; one indicating a certain degree of land use and one indicating pristine nature.

Nevertheless, politically, nationalists often have the greater political draw at the ballot box than do 'Greens'. The result is two-fold. Nationalists will continue to 'mine' environmental discourses for the sake of further legitimating their politics across a wider segment of the electorate. In addition, the discursive overlaps between nationalists and Greens can lead to electoral alliances between the two political movements. Thus, it is no surprise that the Latvian Green Party has had its greatest political moments in conjunction with the Latvian Farmers' Party.

Baltic environmentalism then and now

Environmental movements in the late Soviet period set the process of disintegration in motion. While environmentalism was vital to the Baltic independence movements, a general 'Green' consciousness as seen through political debates, discourses and decision-making has been lacking in post-Soviet Latvia. Thus, we ask, despite active environmental movements in the late Soviet period, why have we seen a depoliticization of environmentalism? First, to see where Latvian environmentalism is today, we have to see the path to this point.

What role did Latvian environmentalism play in the late Soviet period? The Soviet authorities allowed no collective dissent against the state or its policies (see Hough & Fainsod 1979). Some open dissent did develop in the late 1950s, especially concerning the change in bilingual education laws in 1957–1958; however, this dissent was not wide-spread, involved only high-level Latvian political elites, and was promptly dealt with by the Soviet authorities (Bilinsky 1962). The Leonid Brezhnev period led to a tightening by the security services on any act of dissent. Schwartz illustrates how Soviet occupation and ideology attempted to decouple nature from national identity, not having any room for the latter in a Soviet society. She illustrates that the state of the environment slowly worked its way back into the Latvian national discourse from the late 1960s onwards. In the mid-1970s, the Great Tree Liberation Movement set out to protect trees from the onslaughts of collective farm directors. Trees are especially important symbols in Latvian national identity and appear on the face of contemporary banknotes as one would ordinarily expect to see a historical figure. Schwarz states that the Great Tree Liberation Movement set out 'to restore order to the garden and reinstate the exiled Latvian as steward of his individual farm' (2006, p. 61). Importantly, the movement set in motion events that would eventually restore the Latvian nation.

The rise to power of Mikhail Gorbachev in 1985 was accompanied by a fundamental change in state–society relations. While it was still taboo to challenge the legitimacy of the Soviet state, there was room for collective dissent against specific problems, especially if they could be couched in Soviet political discourse. Environmentalism was one such problem. The Baltic environmental movements were mirrored by a new consciousness within the Soviet administrative system of environmental degradation within the Soviet Union. David J. Galbreath (2005, pp. 94–6) illustrates how the Baltic environmental movement became a test run for future collective action that more directly challenged the legitimacy of the state. Effectively, the Latvian environmental movement established a 'repertoire of contention' that would eventually challenge Soviet occupation (on the concept, see Tilly 1995).

The Soviet Latvian environmental movement was primarily focused on the proposed dam on the Daugava River, an artery that cuts Latvia in half, originating in the Russian Federation. Juris Dreifelds (1989, p. 11) illustrates how the larger Soviet environmental consciousness sparked the cancellation of plans to divert two Siberian rivers, the Ob and Enisei, to Central Asia in 1986. At the same time, the Soviet Politburo cancelled a planned canal that would link the Danube and Dnieper Rivers. Yet at the time a new hydroelectric dam was planned for the Daugava. Dreifelds (1989, pp. 13–16) contrasts this late initiative with the 1958 plan that eventually led to the erection of a dam on the Daugava. Note the overlap in time between the 1958 dam proposal and the 1958 bilingual education changes. A first dam (the *Ķegums*) was erected on the Daugava in 1937, followed by the Soviet planned dam in 1958. The 1958 plan encountered opposition from Latvian Communist Party officials, who wanted to see the dams located lower down the river, arguably causing less damage to the scenery and environment. Dreifelds (1989, p. 13) points out that Vēra Kacena, the editor of the journal *Literatūra un Māksla*, was the first to raise public opposition to the 1958 plans. Core to this argument was the rhetoric of archaeology and history, rather than environmentalism, perhaps not surprising considering the embryonic state of the environmental movement in other industrialized societies at the time. After a rebuke in the official party publications, such as the party's paper *Cīņa*, all public discussion of the plan ended. The dam was built in 1967.

The decision to cancel the plans to divert the Ob and the Enisei sparked public opposition to new plans to build a hydroelectric dam on the Daugava in the 1980s. Once again, it was the journal *Literatūra un Māksla* that began the debate with an article in 1986 (1989, p. 16). The article was written by Dainis Īvāns and Artūrs Snips. As in 1958, they used the rhetoric of nation and history, but also employed an ecological argument including issues such as silting, eutrophication (removal of oxygen from the water), fish loss, and flooding. After Īvāns and Snips, the student paper *Padomju Students* began a campaign to oppose the construction of a third dam on the Daugava. According to Dreifelds, the students began collecting signatures throughout Riga. With growing local opposition, the Latvian Council of Ministers ordered a review of the ecological impact of the planned Daugava dam. However, this dam construction was an all-Union affair, meaning that the plans lay with decision-makers in Moscow, not Riga. Thus, we see a clear tension between the center and periphery in the late Soviet period. At the request of the Latvian Council of Ministers, a special commission consisting of members from the Latvian Academy of Sciences, the State Agro-Industrial Committee, and the State Construction Committee performed the first (ever) ecological study of the Daugava dam project. As result of this study and the ensuing debate, the all-Union Academy of Sciences–Biosphere Problems Council passed a resolution in 1987 stating that the planned dam was 'ecologically, economically and socially unsuitable' (1989, p. 19). The USSR Council of Ministers canceled the project on 7 November 1987, only a year after the first signs of public discussion. The Latvian environmental movement coalesced after the victory into the Environmental Protection Club (*Vides Aizsardzības Klubs*, VAK).

The environmental and cultural conservation movements gave way to outright nationalism in 1988. The nationalist movement used many of the same repertoires of the environmental movement: public debate, gathering signatures, organizing committees, and coming together for public displays of opposition. In 1988, both the Latvian People's Front (*Latvijas Tautas Fronte*, LTF) and the Latvian National Independence Movement (*Latvijas Nacionālās Neatkarības Kustība*, LNNK) were created with initially diverging agendas. The LTF initially focused on greater autonomy within the Soviet Union, while the LNNK focused squarely on independence from the USSR. By 1990, both the LTF and the LNNK were calling for independence from the Soviet Union. On 4 May 1990, the Latvian Supreme Soviet declared the restoration of independence and on 21 August 1991 Latvia regained its independence. Thereafter, any environmental discourse was eclipsed by the debates over economic transformation, democratic transition, and the large, non-citizen, Russian-speaking community in Latvia.

In so far as an environmental movement has existed in the post-Soviet period it has primarily focused on the tension between conservation and biodiversity. This tension lies between Western environmental narratives and East European nationalism.

Latvia began efforts to rebuild an agricultural industry while the 'West' began to seek to protect special areas of natural resources. With these two competing narratives, Schwartz offers four case studies that exemplify the tensions between global ecological and local environmental narratives. First, she examines the debates over the Gauja national park in Vidzeme and the Slitere nature reserve in Kurzeme. The Gauja national park was established in the 1970s to protect a particularly Latvian (but not necessarily un-Soviet) agrarian ethnoscape. Within the Slitere nature reserve in northern Kurzeme, there was an attempt to resurrect an economically deprived community by creating a national park. In both cases, there was significant disagreement between the locals, who wanted to maintain a productivist agrarian relationship with the land, while Western advisors attempted to create a park of isolated biodiversity. The narrative of nation and history once again was used to confront a planned rezoning of the land, although this time against an otherwise ecological argument for biodiversity. The productivist agrarian narrative won out over that of the ecological.

Second, the author examines the regeneration efforts around Lake Pape in southwest Latvia near the Lithuanian border. The Lake Pape project was meant to bring socio-economic development to a place of existing natural beauty and importance. With a series of diverse ecosystems, including bogs and wetlands which support an important bird migration route from north to south, the Lake Pape project was an ideal area for ecological protection and conservation. Unlike in Gauja and Slitere, the nationalist productivist agrarian approach did not initially compete with the ecological approach of biodiversity. Rather, the project funders, the Danish Ministry of the Environment, and the project implementation team, WWF-Denmark and WWF-Latvia, began with a dialogue of economic sustainability and biodiversity enhancement. However, a conflict arose when there was a Dutch plan to use the reserve to introduce an ancient breed of horse (Konik horses). This tension illustrated a split between two domestic groups: WWF-Latvia and the Latvian Fund for Nature. Schwartz (2006, pp. 159–61) illustrates how WWF-Latvia attempted to use the Lake Pape project essentially to normalize (within a Western context) Latvian society's relationship with nature (i.e. a post-productivist approach). On the other hand, the Latvian Fund for Nature declared the introduction of horses to be anti-agrarian and anti-national. By 2002, the Lake Pape project was firmly devoted to post-productivist ecological protection.

Finally, Schwarz investigates the debates over forestry management and logging (see also Lazdinis *et al.* 2004; Schwartz 1999). Again, WWF-Latvia found itself up against another domestic group promoting a productivist relationship with nature. Schwartz illustrates how WWF-Latvia attempted to bring about a sustainable logging program in Latvia. This program was strongly resisted by the traditional farming community. WWF-Latvia relied on a 'Swedish method' of sustainability, while the traditional foresters declared that a Latvian tradition of forestry management, born out of Soviet forestry management practices, was the best way to protect Latvia's forests. Schwartz (2006, p. 182) argues that behind the traditional approach was not only a form of Latvian nationalism but an entrenched xenophobia of things Western. In the end, the government, partly pulled along by substantially large Swedish logging companies (e.g. Swedwood, an IKEA company), agreed to implement Western sustainability methods.

Like the other studies, Schwartz's cases are even more important because they were strongly affected by a globalizing method of biodiversity. In the Western context, this can be seen as a normalization of environmentalism in post-Soviet Latvia, exemplified by ecological, grassroots organizations such as VAK and the national chapter of the Friends of the Earth. So far, we have seen where environmentalism and nationalism overlap and fundamentally where they differ. Schwartz's case studies give us an indication of the ecological debates that have been occurring in Latvia since the Soviet period. By looking more closely at political actors, we can see again where environmentalism and nationalism come together with special interests to offer a multicolored Green Party.

The Latvian party system

Independent political movements began forming in Latvia in the late 1980s as a result of the gradual democratization of the Soviet Union. A second period of political organization formation, with many of these newly formed movements choosing to call themselves parties, began in the early 1990s in the run-up to the first post-communist parliamentary elections in 1993. There were three organizational and ideological starting points: (i) the radical nationalist wing of the Latvian independence movement that was composed of the *Pilsoņu Komiteja* (Citizens' Congress) and the LNNK; (ii) the moderate LTF, a broad umbrella grouping that included pro-independence ex-communists as well as moderate Latvian nationalists; and (iii) the reactionary pro-Soviet and anti-independence Latvian Communist Party (*Latvijas Komunistiskā Partija*, LKP) and hard-line Russian-speaker *Interfront* movement.

The contemporary Latvian party system still resembles this spectrum, with parties grouped into Latvian nationalist, centrist/moderate and pro-Russian speaking fractions. This also reflects the central characteristic of the Latvian party system – the cleavage between ethnic Latvians and Russian-speakers. All government coalitions between 1993 and 2009 have been exclusively composed of Latvian parties. As a result, ethnic Russian-speaking parties form a permanent opposition, holding around one-quarter of the parliamentary seats (roughly equivalent to the proportion of ethnic Russian citizens).

Institutions structure and shape both parties and the party system. In the case of Latvia, the legislation governing political party financing and campaigning is particularly salient. The laws on political advertising are very liberal. There are no limits on television, radio or newspaper advertising. Adverts can run right up to and including the day of the election. The only restriction is on political advertising appearing within 100 meters of a polling station on polling day.

The Latvian political party financing law passed in 1995 allowed for annual individual and corporate donations of up to 25,000 lats a year, but banned donations from foreign or unidentified sources.[1] In the summer of 2002 the limit was slimmed down to 10,000 lats. However, revisions to the Law on the Financing of Political Organizations in August 2008 changed the method by which donation limits are calculated to a multiple of the minimum wage; thus the annual donation limit was fixed at 100 minimum salaries (180 lats at the end of 2008), amounting to 18,000 lats in 2009.[2] Parties can also earn money from private enterprise and membership fees, although these are not major sources of revenue. Rather, large individual and corporate donations have made up the bulk of party income.

The August 2008 revisions also made further changes to party election campaigning expenses. Following the 2002 election, the expense of which had led to a popular backlash against political parties, party campaign spending was limited at 20 santimes per voter (thus a maximum of 279,053 lats in the 2006 parliamentary election). In the run-up to the 2005 local government elections, parties identified several loopholes in this law. For example, individual candidates were not affected by spending limits, and could thus spend freely. Moreover, in the 2006 election parties made use of third-party financing as a way of skirting around spending limits. Thus four parties spent more money than the law allowed in the 2006 parliamentary election campaign (including the Green Party, in its coalition with the Farmers Union) (Čigāne 2007). The August 2008 revisions raised the spending cap to around 474,000 lats if an election were to take place in 2009 (Council of Europe: Groups of States Against Corruption 2008).

Modern political campaigning is increasingly expensive, with a myriad of old and new media available to political parties. Inevitably, Latvian parties spend a lot of money on elections. The expense of campaigning has, in turn, led political parties to increasingly rely on wealthy individuals and corporations to fund their activities. This has created the opportunity for wealthy individuals or corporations to capture the political process, as they demand a return on their political contribution/financial investment.

There is a positive correlation between the amount of money spent on campaigning and the number of seats won. Prior to 1998 there were no effective laws regulating party income or expenditure. The first accurate data on party spending are available for the 2002 parliamentary election, when civil society organizations had acquired both the financial and human resources needed to accurately audit campaign spending. Party spending was very high, with parties declaring total spending of 5.3 million lats, although an independent report argued that the real sum may have been up to 1.5 million lats higher (Čigāne 2003). Thus parties spent a total of at least 2.8 lats per voter (rising to 3.9 lats if the additional undeclared figure is included), compared to 1.7 lats in the 2000 American presidential election or 0.4 lats in the 2001 UK parliamentary election (two countries that also have few limits on party overall income and spending) (Čigāne 2003). The Peoples Party (*Tautas Partija*, TP) spent over 1.2 million lats, more than double the spending of any other party. Five hundred thousand lats, or some 10% of all funds raised, came from untraceable sources (Čigāne 2003). Moreover, in 2004 the Latvian anti-corruption bureau reported that 65% of the donations made to parties by individuals in the run-up to the 2002 election were classified as 'dubious', because the individual donors did not have an income commensurate to their donation (Korupcijas Novēršanas un Apkarošanas Birojs 2004).[3]

The party-financing model largely shapes both Latvian parties and the party system. It creates party dependency on corporate groupings and wealthy individuals to compete in both local and parliamentary elections. Indeed, politicians have freely conceded the overarching importance of money in Latvian politics. Dzintars Rasnačs (former Justice Minister and long-standing parliamentary deputy of the For Fatherland

and Freedom/Latvian National Independence Movement Party (*Tevzemei un Brivibai/ LNNK*, TB/LNNK)) has pointed out that 'the party financing model means parties rely on corporate sponsors and individual *éminence grises*, which leads to parties serving the interests of their sponsors, not society as a whole'.[4]

There is little opposition to these practices within the parties themselves, which tend to have small memberships. Party membership in post-communist Europe as a whole has been largely neglected by the leaders, partly due to a perception that 'the practice of liberal democracy... operate(s) at national level and within the narrow confines of the political elite', and partly because party members are of less significance in terms of both party financing and campaigning activities (Lewis 2000, p. 104). Indeed, in contrast to the financial dimension, there is little correlation between party membership and electoral success. New Era (*Jaunais Laiks*, JL), the party winning the largest share of the vote in the 2002 election, had the smallest number of registered members of any party competing in the election, while the Latvian Social Democratic Worker's Party (*Latvijas Sociāldemokrātiskās Strādnieku Partija*, LSDSP), the party with the second-largest membership, failed to pass the 5% threshold for getting into parliament (Auers & Ikstens 2005). Indeed, parties may well wish to make it difficult for new members to join in order to remain small and avoid 'alien takeovers' (Smith-Silivertsen 2004, p. 239).

This combination of a dominant ethnic cleavage, low membership and virtually unlimited media campaigning, as well as a party financing system that relies exclusively on private donations, has led to the capture of political parties by economic interests. The development of the Latvian Green Party over the past decade is a representative case study of this process.

The birth of the Latvian Green Party

As discussed earlier, environmental issues played an important role in the early development of Latvian political discourse in the 1980s. Latvian Green activists first rose to domestic prominence in the 1980s, initially through modest action to preserve Latvia's rural inheritance, then later organizing major protests about the building of a new hydroelectric dam, followed by organizing opposition to the possible construction of an underground metro system in the capital Rīga. The latter actions were organized by VAK, the first organized Green movement in Latvia. Juris Dreifelds quotes Dainis Ivāns, the first leader of the LTF, as stating that 'in Latvia everything began with the movement to save the environment' (Dreifelds 1996, p. 54). Indeed, VAK was one of the founding organizations of the Latvian People's Front, and many of its members were also founding members of the nationalist LNNK (Dreifelds 1996, p. 59).

As a result, the Green movement in Latvia is associated with the drive for independence and with ethnic Latvian, rather than Russian-speaking or cross-cutting, ethnic parties. As Aldis Purs and Artis Pabriks (2001, p. 51) have pointed out, the VAK logo was simply the Latvian flag with Green, rather than red, coloring. And in black and white print, as was most common in late 1980s Latvia, it was essentially indistinguishable from the Latvian flag.

The Latvian Green Party was founded on 13 January 1990, by members of VAK and the LTF. As with all Latvian political parties, the Greens began small, and have remained small. The party initially had 123 members; membership hovered around 200–300 in the 1990s (parties require a minimum of 200 members to be legally registered), and it now claims a membership of around 600 (Latvian Green Party 2008). At the same time, the party charges no membership fee, and initially attracted few financial sponsors (Latvian Green Party 2008).

As a result, the party was forced into taking on coalition partners to share (or, indeed, almost wholly take on) the significant financial burden of political campaigning. Only in the 1993 parliamentary election did the Green Party campaign alone, winning just 1.19% of the vote, well below the 4% (5% in following elections) threshold needed to gain entry to parliament. In the 1995 election the Greens formed an electoral alliance with the nationalist LNNK (which had been partially formed by former members of the VAK in the late 1980s), winning eight seats (6.33% of the vote) in the 100-member Latvian legislature. Half of the elected deputies were members of the Green Party. In the following election the Green Party was forced to seek new partners as LNNK had merged with its fellow Latvian nationalist party, For Fatherland Freedom (*Tevzemei un Brīvībai*, TB). Thus in the 1998 election the Greens formed a coalition with the Christian Democratic Union (*Kristīgi Demokrātiskā Savienība*, KDS), and the Labour Party (*Darba Partija*, DP), but won just 2.29% of the vote, and failed to enter parliament. This history of seamlessly moving from an electoral coalition with nationalists to partnership with a center-left and Christian democratic party reveals the ideological flexibility and political opportunism of the Green Party in Latvia. Indeed, in the 2002 and 2006 elections the Greens formed a successful coalition with the nationalist-agrarian Farmers Union (*Zemnieku Savienība*, ZS). The ZZS coalition won 9.4% of the vote and 12 seats in 2002, and enlarged this to a 16.71% share of the electorate and 18 seats (the second-highest share of the vote in the election) in 2006.

Brown and black: nationalism and the 'Ventspils' oil transit lobby

The two key elements to the post-2002 electoral success of ZZS are nationalist rhetoric and the continuous financial support of the Ventspils oil transit group. The 'brown', or nationalist, dimension has been a part of the Green Party since its inception in 1990, and can be traced back to the origins of the Green movement in Latvia as a proxy for nationalist sentiment. However, the corporate dimension of the party emerged only after its electoral alliance with the Farmers Union in the run-up to the 2002 parliamentary election.

The nationalist rhetoric of ZZS is based on two basic positions: first, support of Latvians as the core ethnic nation in the Latvian territory; second, opposition to modern liberal values, particularly sexual minority rights, which are presented as a foreign-led attack on traditional Latvian values. Both positions serve the same function. An enemy (initially the European Union, but later modified to George Soros and affiliated local organizations such as the Soros Foundation), attacking Latvian national and traditional values, is identified. Having created an external threat, the party then reassures voters that it will protect their national interests and values during this period of globalization, Europeanization, and radical change. The anti-Soros/anti-Western rhetoric also served to deflect attention away from the charges of corruption and bribery laid against ZZS's candidate for prime minister in the 2006 parliamentary election, Aivars Lembergs, by dismissing the charges as a Soros-led plot seeking to undermine Latvian sovereignity (Antonevics 2006).[5] Thus in the campaign leading up to the 2006 parliamentary election, ZZS portrayed itself as a party defending not just Latvian values, but Latvian sovereignty and independence. Environmental discourse only featured in terms of identifying an idyllic rural Latvia with the traditions and values that ZZS aimed to defend from liberal external influence. The centrality of this nationalist, anti-liberal position in the party's identity can be illustrated with a few examples.

The first, nationalist, position can be identified in both the 2002 and 2006 parliamentary election party programs, which emphasized the importance of the Latvian language and the Latvian nation: 'Latvia will be made into a national, beautiful, and powerful state, with Latvian as the only national language and Latvian culture dominant. We believe that only the Latvian nation has the right to determine Latvia's future' (ZZS 2006 Parliamentary Election Program). Moreover, this nationalist rhetoric was reinforced by the prominent campaign roles allocated to more nationalist-oriented politicians recently recruited by the Green Party. Leopolds Ozoliņš, a former hardline nationalist parliamentary deputy of the TB party (from 1993 to 1995), joined the Green Party in the run-up to the 2002 parliamentary election, and was elected to parliament in both 2002 and 2006. He was joined in 2006 by fellow recent Green Party recruit Visvaldis Lācis, an octogenarian self-publishing radical nationalist historian. Both consistently opposed the ratification of Latvia's border treaty with Russia in 2007, partly because it would legalize the handover of part of inter-war Latvia's territory, and partly because it was not accompanied by any Russian acknowledgement, or apology, for the Soviet occupation of Latvia (Krūmiņa 2007). Indeed, in April 2007 Lācis gave a speech in the Latvian parliament that included singing a Latvian SS Legionnaire Second World War song.

The second core position of hostility to modern liberal values is reflected in political speeches rather than in official programmatic or policy positions. Thus it enters the local, Latvian, discourse, but has little resonance beyond Latvia's borders. Programmatically, the Greens appear to be a mainstream European Green party, but their political discourse is far more extreme. One piece of research on homophobic speech in the Latvian parliament found Leopolds Ozoliņš to be an outspoken opponent of gay rights: 'Pederasts are not just content to *make love* in their own bedrooms, they are now shamelessly and openly pushing their perverse lifestyle propaganda on all of us' (Mozaika 2007). Moreover, one of the founders and most prominent members of the Green Party, Arvīds Ulme, as well as Visvaldis Lācis, have both participated in anti-gay-pride demonstrations and have been outspoken in opposing gay rights in Latvia.

This anti-liberal discourse has been accompanied by a rhetoric that seeks to identify ZZS as a national bulwark against the liberal, cosmopolitan and thus anti-Latvian George Soros network that is attempting to capture the Latvian state. Thus the ZZS candidate for prime minister in the 2006 parliamentary election, Aivars Lembergs, has positioned himself as the anti-Soros opposition. He declared that 'Soros has created a very powerful, influential and generously financed network in Latvia, influencing society, law makers and the security services' (LETA 2005). As a result, Lembergs declared the Latvian state to be a 'Soros subsidiary' (LETA 2005), grounding the anti-liberal discourse in a basic nationalist appeal.

This radical rhetoric has been accompanied by expensive media-based election campaigns. The funding base for the 2002 and 2006 election campaign came from the Ventspils transit group. Ventspils is a small Latvian coastal town that serves as the endpoint for an oil pipeline stretching across Latvia from Russia. Before the Russian government cut off the flow of oil in the pipeline to Ventspils, it was one of the largest oil terminals in Europe, transiting over 10% of all Russian oil and oil products. It still transits millions of tonnes of oil via an overland train route from Russia. As one of the few cash-rich Latvian businesses in the 1990s, the interests of the Ventspils group rapidly expanded into the media (the Ventspils group controls two of the three major Latvian daily newspapers – *Latvijas Avize* and *Neatkariga Rita Avize*) and politics.

Ventspils transit-related companies were privatized between 1993 and 1995, when the Latvia's Way (*Latvijas Ceļš*, LC) party dominated the government, including the prime minister's office, the finance ministry and the transport ministry, as well as the privatization bureau (it had won a record 36 out of 100 parliamentary deputies in the 1993 election). LC was the first major party to benefit from the financial largesse of the Ventspils group. Indeed, during the premiership of LC's Vilis Krištopāns in 1998 and 1999, Aivars Lembergs even attended meetings of the Cabinet of Ministers, and the Krištopāns government was nicknamed Lembergs's 'pocket government' (Lapsa & Jančevska 2006, p. 133). After being fired in 1999, Krištopāns' Economics Minister, Ainārs Šlesers, stated that 'the real government leader is Aivars Lembergs' (Huang 1999). In the 1990s the Ventspils group was also financially linked with the LSDSP as well as the more nationalist TB/LNNK (Ijabs 2007).[6]

However, electoral support for all three parties waned in the build-up to the 2002 parliamentary election (indeed, LC and the LSDSP failed to pass the 5% threshold in the 2002 election, while TB/LNNK won just 5.4% of the vote, equating to seven deputies). As a result, the Ventspils group switched its financial support to the ZZS. This can certainly be seen in the swollen financial coffers of the ZZS in 2002 (both parties had struggled for financing in the previous 1998 parliamentary election, and as a result failed to pass the 5% threshold to be elected to parliament).[7] Indeed, the Green Party was already friendly with the Ventspils group in the late 1990s, actively opposing the construction of the Butinge oil terminal in Lithuania, a clear rival to the Ventspils oil transit business, organizing a hunger strike in front of the Lithuanian Embassy in Riga, as well as occupying part of the proposed terminal area in Lithuania. At the same time, it never challenged the ecological impact of the Ventspils terminal, and even went to the extent of inviting Lithuanian Greens to a fact-finding tour of Ventspils (Bierande 1999). The Green Party also had a representative on the board of the Ventspils oil transit company. By 1999 it had already earned the popular nickname '*zaļie-naftinieki*' (Green-oilmen). Thus by the time of the 2002 election the Green Party had already been effectively taken over by the Ventspils transit group.

The Green Party was attractive to the Ventspils group for two reasons. First, the party was small, with few committed, active members, and thus firmly in the control of its leadership. Just 90 people attended the party congress in

February 2008 (LETA 2008). This made a 'takeover' relatively straightforward. Moreover, this meant that, as with the Farmers Union, non-party 'charismatic' candidates could be leveraged onto the ballot. The result, in both 2002 and 2006, has been a mix of extreme nationalists (such as Leopolds Ozoliņš and Visvaldis Lācis), experienced politicians (former LC Prime Minister Vilis Krištopāns), callow, but popular, Olympic medal-winning weightlifters (Viktors Ščerbatihs), as well as former Ventspils group business affiliates (former PriceWaterhouseCoopers accountant Ingrīda Ūdre).

Second, the Greens, in coalition with the Farmers Union, had a well-established political identity – a core political environmental/nationalist image that could be used as the basis for a refined media campaign appealing to ethnic Latvian voters' attachment to rural areas (hence the recruitment of hardline nationalists). The media campaign featured the newly recruited 'charismatic candidates' filmed in rural idylls – such as Ingrīda Ūdre leading a white horse through a green meadow. There was also a grassroots campaign that organized public meetings with candidates followed by subsidized beer festivals, which attracted a substantial turnout in rural areas.

As with the Ventspils group's previous sponsorship of political parties, the business–political relationship is opaque. However, the deep links between the two came into the open in 2006 when the Ventspils group's vehicle for the local elections, the For Ventspils and Latvia Party (*Ventspilī un Latvijai*, V&L), joined the ZZS electoral coalition. The three parties then jointly nominated Aivars Lembergs as their prime ministerial candidate for the upcoming election (ironically, on the same day that the Latvian prosecutor general named him as a suspect in a political corruption and money-laundering investigation) (LETA 2006). He remained their official candidate even after he was subsequently arrested and charged with paying over five million lats in bribes to Latvian political parties and politicians in spring 2007 (Rulle 2008). The prosecutors' investigation also charged that individual deputies were financed in addition to political parties as organizations – the Latvian media has long called the suspected deputies Lembergs's stipendiates (*Lemberga stipendiati*). Further scandal hit the Green Party in fall 2007 with the resignation of Indulis Emsis (the only Green prime minister in European history, and then still co-chairman of the Green Party) from the post of speaker of the Latvian parliament, after he admitted lying to police about the size of a wad of cash (subsequently proved to be $10,000) he had mislaid in the Latvian government's Cabinet Office (LETA 2007).

Since 2002 ZZS has become established as a party of power, being a member of all six government coalitions between 2002 and May 2009. ZZS controlled both the Ministry of Agriculture and the Ministry of Environment in this period, thus ostensibly supporting the interests of their core farm and environmental constituency. At the same time, however, financial sponsorship from the Ventspils Group has led to the Greens adopting a core nationalist anti-internationalist and anti-liberal discourse far from the European Green mainstream.

Conclusion

This essay sought to answer several questions. First, despite active environmental movements in the late Soviet period, why have we seen a depoliticization of environmentalism? We illustrated that the environmentalism discourse of post-Soviet Latvia was both diversified and obscured by the tensions of economic, political and social transition. Second, if environmentalism is not a salient issue in Latvia any longer, what is the nature of contemporary environmental attitudes? This essay illustrated how the link between environmentalism and nationalism developed in the late Soviet period and continued to be represented in recent election campaigning by ZZS and the radical rhetoric of individual deputies, if not actually in the written words of its formal party program. Finally, and most importantly, how has the Green Party managed to remain an important actor in post-Soviet Latvian politics, even to the point of controlling government? We suggest that the explanation can be found in the institutional rules governing the financing and campaigning of political parties in Latvia, which have led the Green Party to adopt a variety of pragmatic electoral alliances with ideologically diverse parties, develop a populist nationalist anti-liberal rhetoric, and rely on party financing from narrow 'black' economic interests. The result is a Green Party in Latvia that is ideologically and programmatically far from the European Green mainstream.

Notes

1 1 lat = 0.7 euro.

2 Law on the Financing of Political Organizations, amended August 2008, 'Politisko Partiju Finansēšanas likums' (available at: http://www.likumi.lv/doc.php?id=36189&from=off, accessed 1 June 2009).

3 For example, the Latvian Anti-Corruption Bureau (KNAB) found that 93% of donations to the Green/Farmers Union were made by people who claimed that the money came from money earned more than three years previously – and the Latvian taxation authorities can only investigate individual tax-returns with a three-year backlog. Thus, the truth of these claims cannot be investigated (Korupcijas Novēršanas un Apkarošanas Birojs 2004).

4 Daunis Auers's interview with D. Rasnačs (TB/LNNK Parliamentary Deputy 1995–present, Minister of Justice 1995–1998), 11 March 2002.

5 Aivars Lembergs, the long-standing Mayor of Ventspils and best-known advocate of the Ventspils group of companies, has been charged by Latvian prosecutors as the *de facto* owner of the offshore businesses that own the major transit-related businesses in Ventspils, and thus the *de facto* leader of the Ventspils group (Lapsa & Jančevska 2006).

6 Parliamentary deputies from all three parties have been interviewed as suspects in the investigation of the 'Lembergs's stipendiates' list discussed later in this section.

7 The total declared income of ZZS in 2002 was 503,292.32 lats (Latvian Anti-Corruption Bureau 2002).

References

Antonevics, Ā. (2006) 'Indulis Emsis: Valsts nozadzēji jāmeklē citā vietā' ['Indulis Emsis: Look elsewhere for robbers of the state'], *Lauku Avize*, 25 July.

Auers, D. & Ikstens, J. (2005) 'The Democratic Role of Political Parties', *How Democratic is Latvia? An Audit of Democracy* (Rīga, University of Latvia Press).

Bierande, R. (1999) 'Ventspilī tiekas Latvijas un Lietuvas zaļie' ['Latvian and Lithuanian Greens meet in Ventspils'], *Latvijas Vestnesis*, 30 April.

Bilinsky, Y. (1962) 'The Soviet Education Law of 1958–9 and Soviet Nationality Policy', *Soviet Studies*, 14, 2, pp. 138–57.

Council of Europe: Groups of States Against Corruption (2008) *Evaluation Report on Latvia on Transparency of Party Funding*, available at: http://www.coe.int/t/dg1/Greco/evaluations/round3/GrecoEval3(2008)1_Latvia_Two_EN.pdf, accessed 13 January 2009.

Čigāne, L. (2003) *Analysis of party spending in the 8th Saeima elections* (Rīga, Soros Foundation and Transparency International).

Čigāne, L. (2007) *2006.gada Saeimas vēlēšanans medijos: partiju reprezentācija un slēptā reklama* (Riga, Providus).

Dreifelds, J. (1989) 'Two Latvian Dams: Two Confrontations', *Baltic Forum*, 6, 1, pp. 11–24.

Dreifelds, J. (1996) *Latvia in Transition* (Cambridge, Cambridge University Press).

Galbreath, D. J. (2005) *Nation-Building and Minority Politics in Post-Socialist States: Interests, Influence and Identities in Estonia and Latvia* (Stuttgart, Ibidem Verlag).

Hamilton, P. (2002) 'The Greening of Nationalism: Nationalising Nature in Europe', *Environmental Politics*, 11, 2, pp. 27–48.

Huang, M. (1999) 'They're all paid off in cash, top offices, and girls', *Central Europe Review*, 34, p. 17, available at: http://www.ce-review.org/authorarchives/amber_archive/amber34old.html, accessed 1 June 2009.

Hough, J. F. & Fainsod, M. (1979) *How the Soviet Union is Governed* (New York, Harvard University Press).

Ijabs, I. (2007) 'Azkabānas Gūsteknis—Zaļo Zemnieku perspektīvas' ['The prisoner of Azkhaban—from a Green Farmers perspective'], available at: www.politika.lv, accessed 18 March 2008.

Klandermans, B. (1991) 'New Social Movements and Resource Mobilization: The European and the American Revisited', in Rucht, R. (ed.) (1991) *Research on Social Movements: The State of the Art in Western Europe and the USA* (Boulder, CO, Westview Press).

Korupcijas novēršanas un apkarošanas birojs (2004) 'Pretrunā ar likumu partiju saņemtie ziedojumi' ['Illegal donations to parties'], available at: http://www.knab.lv/offences/article.php?id=20597, accessed 15 March 2007.

Kriesi, H., Koopmans, R., Duyvendak, J. W. & Giugni, M. G. (1992) 'New Social Movements and Political Opportunities in Western Europe', *European Journal of Political Research*, 22, 2, pp. 219–44.

Krumina, I. (2007) 'Robezliguma ratifikaciju nesasteigsot' ['Ratification of the Border Treaty will not be hurried'], *Latvijas Avize*, 20 April.

Lapsa, L. & Jančevska, K. (2006) *Kas ir Lembergs?* [*What is Lembergs?*] (Rīga, Pilsētai un Pasaulei).

Larana, E., Johnston, H. & Gusfield, J. R. (1998) *New Social Movements: From Ideology to Identity* (Philadelphia, Temple University Press).

Latvian Green Party (2008) 'Īsa intervija ar Zaļo Partiju'['Short interview with the Green Party'], available at: http://www.zp.lv, accessed 28 February 2008.

Lazdinis, M., Carver, A., Carlsson, L., Tőnnisson, K. & Vilkriste, L. (2004) 'Forest Policy Networks in Changing Political Systems: Case Study of the Baltic States', *Journal of Baltic Studies*, 35, 4, pp. 402–19.

LETA (2005) 'Lembergs: Latvija par ietekmi cinas Soross un Berezovskis' ['Lembergs: Berezovsky and Lembergs vie for influence in Latvia'], *LETA News Agency*, 1 November.

LETA (2006) 'Lembergs nekandidēs vēlēšanās, bet būs ZZS premjera kandidāts' ['Lembergs nominated as Green/Farmers Union prime minister candidate, but will not stand for election to parliament'], *LETA News Agency*, 30 July.

LETA (2007) 'Emsis atkāpjas no Saeimas priekšsēdētāja amata un noliek mandātu' ['Emsis steps down as speaker of parliament, and resigns as a parliamentary deputy'], *LETA News Agency*, 21 September.

LETA (2008) 'Latvijas Zaļās partijas kongress pulcē ap 90 biedru' ['The Latvian Green party congress gathers 90 members'], *LETA News Agency*, 2 February.

Lewis, P. G. (2000) *Political parties in post-communist Europe* (London, Routledge).

Mozaika (2007) *Homofobiskā runa Latvijā: Politiķu Monitorings [Homophobic speech in Latvia: monitoring of politicians]* (Rīga, Mozaika).

Pabriks, A. & Purs, A. (2001) *Latvia: The challenges of change* (London, Routledge).

Rulle, B. (2008) 'Gads pēc apcietinājuma. Kas mainījies? Kas notiks tālāk?' ['A year after his arrest. What has changed? What's going to happen?'], *Diena*, 14 March.

Schwartz, K. Z. S. (1999) 'Nature, Development and National Identity: The battle over Sustainable Forestry in Latvia', *Environmental Politics*, 8, 3, pp. 99–118.

Schwartz, K. Z. S. (2006) *Nature and National Identity after Communism* (Pittsburgh, University of Pittsburgh).

Smith-Silvertsen, H. (2004) 'Why bigger party membership organizations in Lithuania than in Latvia 1995–2000?', *East European Quarterly*, 38, 2, pp. 215–59.

Tilly, C. (1995) 'Contentious Repertoires in Great Britain 1758–1834', in Traugott, M. (ed.) (1995) *Repertoires and Cycles of Collective Action* (Durham, NC & London, Duke University Press).

WITHOUT A TINGE OF RED: THE FALL AND RISE OF THE ESTONIAN GREENS

Allan Sikk and Rune Holmgaard Andersen

In the 2007 Estonian parliamentary election a new Green Party succeeded in winning 7.1% of the vote and returned six members to the 101-strong *Riigikogu*. The party had been registered just three months earlier, but its history goes back to the late 1980s, when the Green movement played an important role in Estonia regaining her independence. The 'Phosphate War' (sometimes called 'the Phosphate Spring') which the Green movement 'fought' against the Soviet authorities' plans for large-scale phosphate mining in north-eastern Estonia has often been considered a crucial step on the path towards independence. The movement tested Gorbachev's commitment to *glasnost'* and increased the limits of what was tolerable under the then regime. However, the Green movement lost its momentum after Estonia became independent and its offshoots soon vanished from the political scene.

In this study we analyze the formative years of the new party with reference to its historical and ideological origins. Specifically, we seek to locate it within the broader picture of European Green politics. The Green parties in Western Europe have usually been linked to the rise of post-material values pertaining to the 'New Left'. Based on the evidence available, we argue that the Estonian Green Party shares little more than a name and a heightened sensitivity to environmental issues with its West European namesakes.

In the first section of the essay we propose three models of new party emergence by contrasting the classical cleavage-based scenario with alternatives that are all relevant to the development of the Estonian Greens. As mentioned above, the Green

Party was not established from scratch and the second part of the study will sketch the historical background of environmental politics in Estonia relevant for understanding the contemporary party. The penultimate section profiles the party's voters and their motivations using survey and ecological data. The concluding section discusses the findings and argues that the Estonian Green Party differentiates itself from its Western namesakes with regard to its political goals, placement on the left–right continuum and many characteristics of its voters. We contend that rather than belonging to the realm of *post-modernism*, the Estonian Green Party might be better described as being *ultra-modernist*.

Models of new party emergence

Even though conventional theory stresses the importance of societal or value change as the main determinant behind the rise of new political parties (e.g. Kitschelt 1988; Lipset & Rokkan 1967; Müller-Rommel 1989), they can emerge for a variety of reasons. Below, we present the traditional cleavage model together with two alternatives – the challenger and the entrepreneurial party. Even though the models are theoretically distinct ideal types, they all help to understand the reasons behind the emergence of the Estonian Greens.

The classic cleavage model

The classic cleavage approach to new parties focuses on changes in social structures or social values as the driving force behind new parties. The great ancestor of this approach is the seminal work by Seymour Martin Lipset and Stein Rokkan (1967). They studied the evolution of West European party systems and linked the emergence of political parties to the surfacing of new political divisions or cleavages. Their study stopped short of looking at Green parties, however. While Lipset and Rokkan analysed at *structural* conflicts in a society, the rise of Green politics has been attributed to a conflict between materialist and post-materialist *values* (Inglehart 1997, 1977; Müller-Rommel 1989).

Yet no deterministic link is to be assumed between the prevalence of post-materialist values and success of Green parties. For example, despite a high prevalence of post-materialist values, a significant Green party has never got off the ground in Denmark (von Beyme 1985, p. 133). Two main reasons can be identified for such failures: existing parties can accommodate the new demands or the electoral system can be hostile to newcomers (Hug 2001; Rohrschneider 1993). Other restrictive institutions – such as stringent party registration requirements and party financing rules favoring incumbents – and a lack of resources available for mobilizing the electorate may also play a part (Lucardie 2000; Sikk, A. 2006). On the other hand, as will be shown later, the rise of post-material values is not a necessary condition for the spread of environmental concerns. Remarkably, very little post-materialism was involved in the rise of ecological movements in the communist countries during the late 1980s. Hence, the salience of environmental issues and post-materialist values need not go hand in hand.

Challenger parties

In terms of the typology proposed by Paul Lucardie (2000), the first model portrays a 'prophetic' party in that it articulates a new ideology. However, it is unreasonable to assume that *all* new parties reflect structural or value changes; in fact, a new party need not voice novel political demands at all. A 'challenger' or 'purifier' party (Lucardie 2000; Rochon 1985) comes to life as old parties representing an ideology or advocating a concern fail in some sense. Some parties may voice political demands that are not directly related to social divides. Such parties vary, ranging from anti-corruption parties that emerged in Estonia and Latvia in the early 2000s to various personalistic and populist parties seen in many countries (see Sikk, A. 2006). The failure of old parties to represent concerns or to deliver policies in a trustworthy way leads to political disenchantment. This in turn expands the pool of voters only waiting to be mobilized, increasing the incentives for new political parties. Note that we make no normative judgments about the preferability of the cleavage model over the challenger party model for the sake of democracy. Arguably, if corruption among the political elite becomes widespread, an anti-establishment party can be a genuinely progressive force (Sikk 2007). Then again, social divides can lead to the rise of parties with poor democratic credentials.

The challenger party model makes weak assumptions regarding party supporters – they want the newcomer to break the status quo. While some may sympathize with the concerns raised by the party, for others the dislike of the incumbents may be the real driving force. Protest votes are vital for programmatically vague parties but they may also be picked up by parties with clear programs. While most voters presumably choose parties according to programmatic affinity, secondary qualities of political parties – such as the likeability of their leaders or candidates – can be very important. In particular, voters who wish to punish incumbents may not pay much attention to minutiae of policy differences.

Entrepreneurial parties

All political parties have an agenda that goes beyond or even against their publicly stated aims. This also goes for highly programmatic parties that may choose to veil policy plans that can be costly in a catch-all electoral environment. Conservatives, for instance, may hesitate to talk about cuts in public services, socialists may refrain from mentioning tax increases, nationalists can tone down their rhetoric towards minorities and rural parties may conceal plans that can increase the cost of food. The Greens may try to avoid mentioning the full economic costs of ecologically sound policies. Some parties have more sinister covert aims – using public office for private gain. This may take a range of forms. In the more benign cases, they try to satisfy party leaders' private needs or mere vanity. In more malign cases, a party may relinquish its representative functions, and turn into something resembling an enterprise, merely producing returns for its founders, as has been argued with regard to Silvio Berlusconi's *Forza Italia* (Hopkin & Paolucci 1999). Such *entrepreneurial* parties still need at least a rudimentary political profile for electoral and legitimizing purposes and have to deliver some real policies when in power. Needless to say, there is a continuum between the improbable extremes of a fully sincere/impeccable and a totally hypocritical/sinister party. Yet, in the world of ideal types, we can still conceive of a party that has been set up by 'entrepreneurs' with the sole aim of serving their private ends. A very similar 'business firm' model of parties has been proposed by Hopkin and Paolucci (1999), but we would like to stress that the ultimate beneficiaries may be party donors rather than a party itself and yet the whole affair may be a positive-sum game – for example, we could imagine a situation where government support for ecological industries benefits the corporate party donors yet at the same time produces positive externalities for the society at large. Mediating factors such as electoral institutions and resources influence the chances of entrepreneurial parties. They are perhaps even more important here than in the first two models, because such parties face higher costs of camouflaging their intentions and they may have to spend more for mobilization purposes since they patently lack a natural constituency.

Classifying the Estonian Green Party may look like an easy task. After all, the West European Green parties essentially developed following the cleavage model; that is, as a consequence of a value change among their respective populations. As shown below, this is only partly true of Estonian Greens. The party manifesto and its leaders emphasize the environmental issue that was previously largely overlooked in Estonian politics. The Greens also stress the need to engage citizens and NGOs more in policy-making that is more easily associated with post-modern values. Their voters stand out as being younger, more educated and in some ways socially more active than the average voter. In this regard they appear similar to the 'post-modernized' support base of the West European Greens. Yet, due to historical context, the Estonian Green Party and its supporters are slightly right-leaning, in stark contrast to their left-of-centre West European counterparts.

The Estonian Greens seem only vaguely post-materialist at best. The party manifesto often appears socially conservative and 'ultra-modernist' with its frequent references to technological innovation as a panacea for environmental degradation under an essentially capitalist economic system. There is not much in the way of a value or social divide between the Greens and other main political parties. Remarkably, we did not even find clear evidence that environmental issues rank high among party supporters' concerns. The party has certainly managed to mobilize genuine environmentalists and some post-modernist voters, but a large segment of its support base has been made up by voters with diverse backgrounds drawn to the party because of its general likeability or out of disillusionment with old political parties. There is also evidence of a symbiosis between political and business entrepreneurs, as the Greens' election campaign was partly sponsored by people active in the renewable energy sector ('Vastused küsimustele' 2007).

A short history of the Estonian Greens

The fortunes of Estonian Greens fluctuated remarkably between 1987 and 2007. In this section, we give a brief historical overview of developments, with particular attention to aspects that have relevance for understanding the success of the Green Party in 2007.[1] Figure 1 shows a timeline of the most important events. We divide the 20-year period into three phases: (1) 1987–1992: the rise of the ecological movement

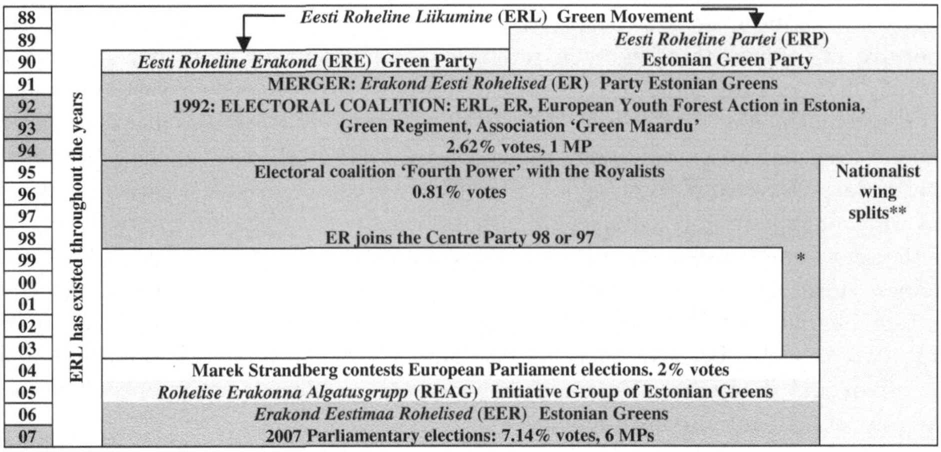

FIGURE 1 Development of Green political organizations in Estonia, 1988–2008. *Notes*: shading in the left-most column indicates years when Greens were represented in the parliament. * In some respects, the Green Party continued to exist 1999–2006 even though it was not officially registered (see text). ** The nationalist wing did not support the electoral coalition with the Royalists and split from the ER (Alvela 2005).

and its transformation into a Green Party in the first post-independence parliament; (2) 1992–2004: the decline of the Green Party into extinction; and (3) 2004–2007: the rebirth of the Green Party and its return to parliament.

Phase 1: 1987–1992

The advent of Green politics in Estonia dates back to 1987 when the Soviet central government started preparations for large-scale phosphate mining in north-eastern Estonia. The plans triggered a widespread reaction among the population.[2] While phosphate mining would have inflicted irreparable damage upon the environment, ethnic identity concerns featured high on the agenda of the emerging Green movement. The Soviet model of industrial development had paid only scant attention to the environment and had been based on the influx of labor from the rest of the Soviet Union, mostly Russia. Even though the immigration was not forced in a strict sense, the people were tempted to move to Estonia by benefits – such as adequate housing – that were often not available for the local population. The newly arrived had few incentives to learn the local language and, consequently, the status of the Estonian language declined while Russian steadily took over in the public sphere. This created a logical connection between policies detrimental to the Estonian natural environment and those that threatened the Estonians' national identity.

The Estonian Green Movement (*Eesti Roheline Liikumine*) was established in April 1988. Its main activities included campaigns to clean up after the Soviet Army (e.g. removal of barbed wire from the north Estonian coast),[3] peaceful demonstrations against the occupying forces and biking expeditions to areas affected by environmental degradation.[4] During its heyday, the Green Movement gained both national importance and international fame (Geyer 1988) when it passed a declaration on

25 August 1988 demanding that Bruno Saul step down from his post as the Chair of Estonian SSR's Council of Ministers within 60 days. Eventually, Saul did step down, though it remains unclear to what extent this was related to the actions of the Green Movement (Toomla 1999, p. 222).

The birth of the Estonian Green Party was as complex as party development often got in Central and Eastern Europe during the late communist and early post-communist years. The Green Movement initially gave rise to two parties – *Eesti Roheline Partei* (ERP) was established in August 1989 and *Eesti Roheline Erakond* (ERE) in October 1990 – both translate into English as 'Estonian Green Party'. By October 1989 ERP had already become a member of the European Greens, but it soon virtually ceased to exist, merging with ERE in December 1991 to form the 'Estonian Greens' (*Erakond 'Eesti Rohelised'*, Toomla 1999, pp. 222–3).

In the 1992 *Riigikogu* elections, the electoral coalition of the Estonian Green Movement (*Eesti Roheline Liikumine*), Estonian Green Party (*Eesti Rohelised*), European Youth Forest Action in Estonia (*Euroopa Noorte Metsaaktsioon Eestis*), Green Regiment (*Roheline Rügement*) and Association 'Green Maardu' (*ühendus 'Roheline Maardu'*) was supported by 2.62% of voters. Although the coalition did not cross the 5% electoral threshold, it did receive a single mandate in the parliament. This was due to the personal popularity of Rein Järlik, who received 9% of the votes in his district, thereby obtaining a personal mandate. The electoral list was highly heterogeneous and many of the coalition's candidates later found themselves representing a variety of other parties. Their top candidate in 1992, Jüri Liim, later contested national and local elections with the Royalists (*Rojalistid*), the Forest Party (*Metsaerakond*), the coalition of the Right and Moderates (*Paremerakonnad ja Mõõdukad*), the Union of Tallinn City Dwellers (*Linnakodanike Liit*), the People's Union (*Rahvaliit*), and the *Pro Patria* Union (*Isamaaliit*). He later became infamous for threatening to blow up the controversial Bronze Soldier in Tallinn.[5] Jüri Martin, who in 1995 succeeded Liim as the Chair of the Party, was the driving force behind its later merger with the Centre Party (*Keskerakond*). Rein Järlik, the only Green MP elected in 1992, joined the Coalition Party before the end of his term. Subsequently, he has been active as the president of the August 20 Club, the organization consisting of the members of the Estonian Supreme Soviet that passed the Declaration of Independence in 1991 (Järlik 2004). Einar Laigna – an ordained Catholic priest and high-ranking military officer, and purportedly a radical theologian (Kärmas 2001) – ran for the 2007 *Riigikogu* elections on the rural People's Union list. In 2003, their list also included Mart Helme, the sixth-ranking candidate of the Green Party in 1992. From 1995 to 1999, Helme served as Estonian ambassador to Russia, establishing himself later as a leading Eurosceptic. In 2004 he was expelled from the People's Union after a row with the party leadership over the removal of a controversial war monument in Lihula. Valdur Lahtvee (ranked number 5 in 1992) and Maret Merisaar (number 12) were both elected to the *Riigikogu* for the Green Party in 2007.

The life histories of the 1992 candidates reveal the heterogeneity among the original Green activists. In this sense, the Green Party (as an outgrowth of the Green Movement) is similar to most other anti-communist movements in Central and Eastern Europe. After their primary objectives had been achieved – in the case of the Estonian Greens national independence, an end to environmentally damaging Soviet policies and the shutting down of many heavy polluters – the parties or movements did not simply call it a day, but split into fractions. Some of the original Greens remained active within the movement and later formed the core of the reincarnated Green Party; some were absorbed by other parties. We contend that such party migration results from the fact that political entrepreneurs may find it difficult to leave the scene after the passing of their organization's *raison d'être*. After losing their original common denominator, they may have little in common. The Green activists followed very different paths – some joining national-conservative parties such as *Pro Patria* or People's Union, and others the center-left and minority-friendly Centre Party.

Phase 2: 1992–2004

As noted earlier, the only Estonian Green MP prior to the 2007 election left the party before the end of his parliamentary term. Meanwhile, the importance of the Estonian Greens in national politics had vanished and in 1995 they joined an electoral coalition with the Royalists, who faced similar problems.[6] The coalition was supported by a meager 0.8% of voters. Joining the Royalists alienated the nationalist wing led by Jüri Liim. Prior to the 1999 parliamentary elections the remainder of the party merged with the Centre Party, losing further members in the process. The formal merger was made necessary as electoral coalitions were banned in 1998 and henceforth only registered political parties could put forward candidate lists.

In some ways, the rump Green Party continued to exist even as it failed to survive as a registered party after falling short of the 1,000 members required by Estonian Political Parties Act (Antsov *et al.* 1998). It continued to be a member of the European Greens, which it had joined in October 1989.[7] The Green Movement was active throughout the two decades. Some Green politicians even saw advantages in operating an NGO as that allowed for more freedom of action – there was less need for the compromises often required of political parties in the parliament (Maret Merisaar, cited in Rudi *et al.* 2000).

Phase 3: 2004–2007

The late 1990s and early 2000s saw significant changes in Estonia's society, economy, politics and international situation. In economic and political terms, the country dramatically changed direction from East to West, joining the EU and NATO in 2004. Economically, Estonia established itself as one of the pioneers of market reforms among the post-communist countries. It witnessed impressive economic growth – between 1999[8] and 2007 Estonian GDP grew on average by 8% annually (EBRD 2008). The booming economy produced a considerable new middle class who did not have to worry obsessively about earning their daily bread. Whereas the middle classes that formed the support base of the West European Green movements and parties were linked to the strong and bureaucratic welfare state, the new middle class in Estonia was born under robust free-market capitalism.

For several reasons, the environment made a comeback in newspaper headlines and on the political agenda. In 2006 three major ecological disasters hit the Gulf of Finland. In January, an oil spill from an unidentified vessel reached Nõva on Estonia's north-western coast. Thousands of wintering seabirds were affected and volunteers turned up in great numbers to help clean up the seashore and those birds with chances of survival. Barely two months later a cargo ship, Runner-4, carrying aluminum sank near the small Estonian island of Vaindloo. It was feared that a large amount of heavy oil would leak into the sea (Soomere & Quak 2007). In the fall, the infamous tanker Probo Koala, which had left toxic waste in several west-African countries, entered the port of Paldiski in north-western Estonia (*Africa Confidential* 2006). Greenpeace actions and press releases from the Green Party initiative group both hit the headlines, once again bringing the environment to the top of the political agenda.

Global hikes in energy prices, fears about the stability of supplies and evidence that Russia was using oil and gas supplies as a political means to influence countries of the former Soviet Union[9] elevated energy to the status of a major political issue. Unsurprisingly, these developments have led many countries to discuss diversification of their energy supplies. In Estonia, the prospects of nuclear energy were vigorously debated. Domestic political turmoil erupted after the three Baltic prime ministers signed a declaration stating the aim of developing a jointly financed nuclear power plant in Lithuania after the closure of Ignalina in 2009 (*ENS News* 2006). The declaration received considerable attention in the Estonian media and was controversial, as there was no hint of such plans in the long-term electricity strategy which the parliament had approved only a few months earlier (Poom 2007). The plans triggered a strong reaction from the Green movement and stirred memories of the Chernobyl catastrophe (Sikk, R. 2006). In July 2007, when the Green Party was already in the parliament, it initiated an extraordinary sitting of the *Riigikogu* after Prime Minister Andrus Ansip signed a new and more specific nuclear energy memorandum with his Baltic colleagues.[10] The future of oil shale mining also became an issue of political debate.[11] Another extensively discussed topic was the plan to build the Nordstream sub-surface gas pipeline directly from Russia to Germany, cutting the Baltic states off from the Russian oil supply chains to Western Europe and creating environmental hazards for the Baltic Sea (Karm 2008). Here, security concerns overlapped with environmental concerns, which echoes the late 1980s in that the Eastern neighbor is seen as a source of danger to both national sovereignty and the environment. Even if oil shale represents the optimal solution to current energy problems from a cost and security point of view – and is presumably popular among the population at large – it is the least appealing from an environmental standpoint. However, the high ranking of these issues on the political agenda gave the Greens a distinct voice in some of the key political debates.

Thus, the increased political salience of environmental issues – coupled with the existence of a potential support base affluent enough to prioritize the environment over material gain – created new possibilities for the Green Party. Other conditions, however, still created substantial hindrances. In order to be registered and to contest parliamentary elections, a party needs at least 1,000 permanent members.[12] Besides the cost of mobilizing members behind a would-be party, the cost of campaigning has been high and is increasing in Estonia – between 1995 and 2007 total officially reported expenditures increased nearly ten-fold from €1.2 to €10.7 per vote cast (Select Committee of Riigikogu . . . n.d.; Sikk, A. 2006, p. 102). This increase clearly exceeds the impressive GDP growth figures reported above. For parliamentary parties, the increase has been partly compensated for by substantial increases in direct

TABLE 1 *Riigikogu* and European Parliament election results

	2003 (RK)		2004 (EP)		2007 (RK)	
	V%	S%	V%	S%	V%	S%
Reform Party	17.7	18.8	12.2	16.6	27.8	30.8
Centre Party	25.4	27.7	17.5	16.6	26.1	28.7
Pro Patria & Res Publica Union	–	–	–	–	17.9	18.8
Moderates/Social Democrats	7.0	5.9	36.8	50.0	10.6	9.9
People's Union	13.0	12.9	8.0	0	7.1	5.9
Greens	–	–	2.3*	0	7.1	5.9
Res Publica	24.6	27.7	6.7	0	–	–
Pro Patria	7.3	6.9	10.5	16.6	–	–
Others	5.0	0	6.0	0	3.4	0
Total	100.0	100.0	100.0	100.0	100.0	100.0

Source: Estonian National Electoral Commission (available at: www.vvk.ee, accessed 23 April 2009).
*Marek Strandberg as independent candidate.

public funding, but state financing for extra-parliamentary parties is small and is only available to parties that won at least 1% of votes in the most recent *Riigikogu* election.

The revival of the Green Party therefore progressed in a series of careful steps. The take-off point was the candidacy in the 2004 European Parliament elections of Marek Strandberg, who later became the main initiator of the new party. He saved the parliamentary parties from an unpleasant surprise by failing to get elected, but he managed to win 2.3% of the vote. His intention to re-establish the Green Party was already publicly known and received an impetus from his decent electoral performance. At the very least he managed to get some free and positive publicity for his persona and his techno-green (i.e. protecting environment by using new technology and supporting R&D) ideas.[13] The Green Party Initiative Group was established as an NGO in May 2005. The membership requirement was finally met in November 2006 and the party – now called the Party of Estonian Greens (*Erakond Eestimaa Rohelised*) – held its inaugural general assembly on 25 November 2006.

While some public opinion surveys before the March 2007 election listed the Greens as the third most popular party in Estonia (Koch 2007), the eventual result was less sensational but still good for a party established barely three months earlier (see Table 1). Following the election the Greens were invited to coalition negotiations with the Reform Party (*Reformierakond*), the *Pro Patria* & *Res Publica* Union (*Isamaa ja Res Publica Liit*) and the Social Democrats (*Sotsiaaldemokraatlik Erakond*) and were dumped only halfway through the talks. However, the Greens were initially satisfied that many of their proposals that had made it into the government manifesto were not dropped with the party.[14]

Profiling the Estonian Greens

In this section, we present a profile of Estonian Greens based on survey and ecological data. Shortly after the 2007 *Riigikogu* election the Department of Political Science, University of Tartu, conducted the Estonian national election survey ($N = 1{,}008$).

The number of Green Party supporters among the respondents was proportional to electoral results and notwithstanding the relatively small number of Green voters (60) the data do provide some fairly robust information about the party's support base. In addition, we have analyzed Green support using a database of municipal voting results and variables from regional statistics databases compiled by Daniel Bochsler and Allan Sikk.[15] Furthermore, we have used the party's programmatic documents, articles and interview findings for help and guidance in interpreting statistical data.

This section is divided into two parts. First, we take a look at socio-demographic characteristics of Green voters together with vote changes between the 2003 and 2007 parliamentary elections. Thereafter, we analyze the political alignment and value structures of the Green voters. Our principal objective is to compare Estonian Greens to their West European counterparts and draw insights from the data that will enable us to locate the Green Party between the models of new party emergence presented at the beginning of this study.

Evidence on whether the Estonian Greens share characteristics with West European ecological parties is mixed. On the one hand, the Estonian Greens advocated more frequent use of direct democracy and empowerment of civil society organizations in decision-making. Initially the party also opted for a flat leadership structure with a number of spokespersons rather than a single leader.[16] On the other hand, the views of its leaders have been supportive of the essentials of the capitalist economic system – the introduction of environmental restrictions is seen as a modern equivalent of the ban on slavery that made capitalism more humane without sounding its death toll ('Vastused küsimustele' 2007). The Greens did not rule out coalitions with any Estonian parties,[17] characterized themselves as 'conservative Greens' ('Vastused küsimustele' 2007) and were initially happy with the prospect of joining a right-of-center cabinet.[18] In this the Estonian party deviates from the general trend among West European Greens, who have clearly been aligned with the left (Mair 2007, p. 215), but similar to the recently successful Czech Greens, who align themselves with the right (Smith 2006, p. 4).

The data from the Estonian election survey reveal that the Green Party received virtually no support among non-Estonians. This could partly be blamed on the fairly nationalist and anti-migration historical backdrop of the ecological movement, even if the relationship between immigration and environmental degradation was manifestly spurious – both were the result of Soviet policies of colonization. However, as it is difficult to find much that could annoy the Russian-speakers in the present-day party, it is equally likely that the lack of minority support results from a thin pocketbook that did not allow the party to run a bilingual campaign. Besides parallel campaign requirements, mobilizing Russian voters was difficult as their support for the (predominantly ethnic Estonian) Centre Party had been rather stable over the last couple of elections.

Our data did not show any significant gender or personal income bias in the support for the Green Party – women were as likely to support the party as men and the average income of Green supporters approached the overall average. However, the support was slightly higher in urban areas, when controlling for the strong effect of ethnicity.[19] The support is rather flat over the age groups, though it drops off among the elderly (see Figure 2). In contrast to the early days of Green parties in

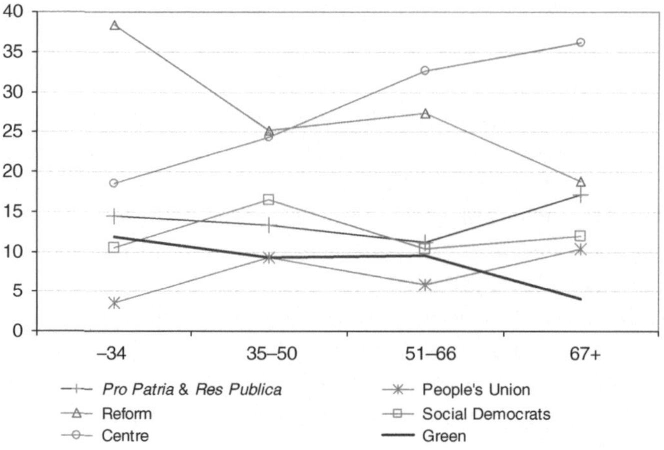

FIGURE 2 Support for political parties by age group (%), *Riigikogu* elections 2007.

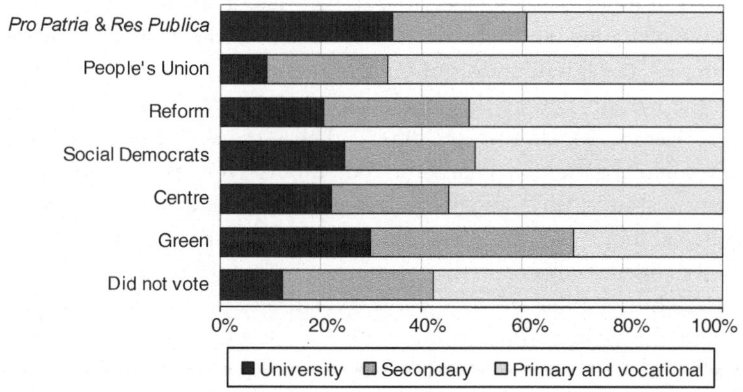

FIGURE 3 Level of education among supporters of political parties, *Riigikogu* elections 2007.

Western Europe, the support is only modestly higher than average among the younger cohorts. On the other hand, similarly to West European ecological parties, the supporters of the Green Party are more educated than those of almost any other political party (see Figure 3) and the average level of education would be even more clearly above all other parties if we included current university students.

Figures 4 and 5 show the sources of Green support in terms of their voters' previous preferences. The Moderates (*Rahvaerakond Mõõdukad*, now the Social Democrats) had the highest defection rate to the Greens (17.0%) among all parties. Also, over 5% of those who supported *Pro Patria*, *Res Publica*, or the Reform Party in 2003 opted for the Greens in 2007. When looking at it from the Greens' point of view, their voters had a rather heterogeneous voting history and, noticeably, more than one in five had abstained in the previous parliamentary elections. Thus, while the

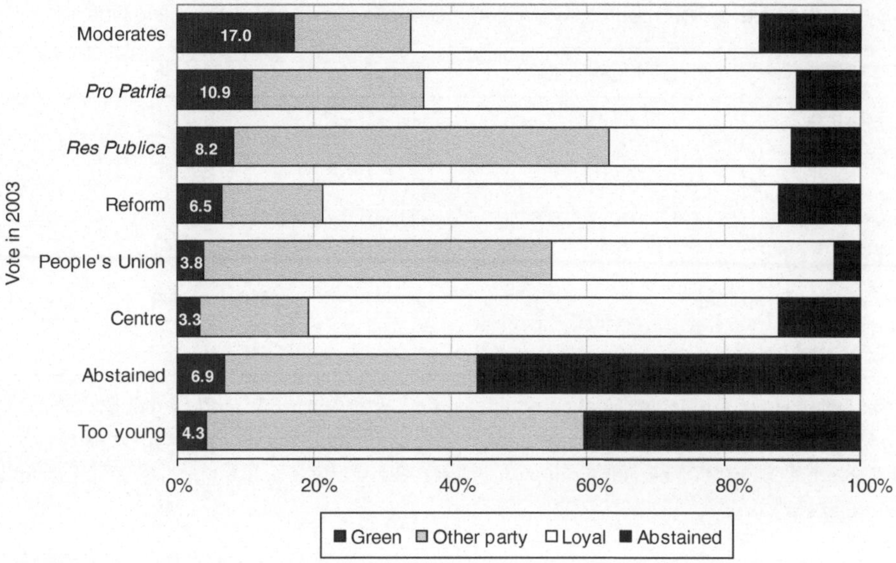

FIGURE 4 Vote switching from 2003 to 2007 *Riigikogu* elections. *Notes*: numbers in white show the proportion of party's voters defecting to the Greens. Before the 2007 election, *Pro Patria* and *Res Publica* merged (both marked as 'loyal' here) and the Moderates were renamed the Social Democratic Party.

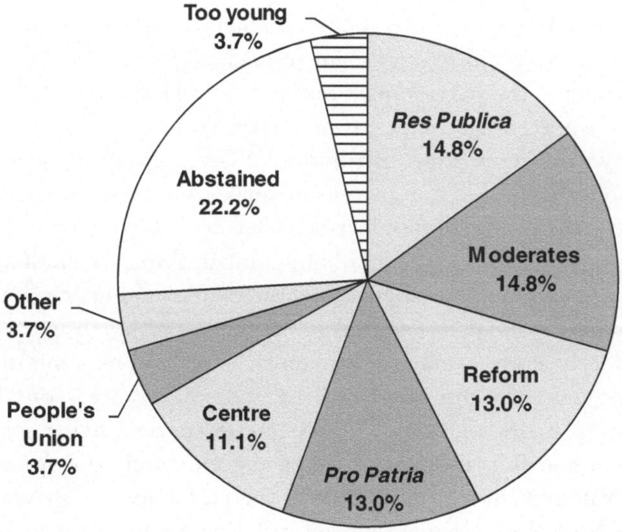

FIGURE 5 2007 Green voters' preference in 2003 *Riigikogu* elections.

Greens managed to mobilize *some* absentees, the bulk of its popularity came at the cost of other parties.

Table 2 shows an OLS regression model of Green Party support across Estonian municipalities based on 2007 electoral data, personal income tax per capita in 2006

TABLE 2 OLS regression model of Green Party support in municipalities

	Mean	Median	Range	b	SE	T	p
Dependent variable							
Green vote%	6.86	7.03	0 to 20				
Independent variables							
Eastern Slav citizens %*	16.7	9.24	0.138 to 91.9	−0.0719	0.00712	−10.10	0.000
ln(income tax per capita 2006)	8.64	8.68	7.47 to 9.72	2.64	0.643	4.10	0.000
Dummy for Strandberg			(0 or 1)	0.898	0.489	1.84	0.067
Constant				−15.1	5.41	−2.79	0.006
$R^2 = 0.547$, N = 234							

Source: Database of municipal election results compiled by Allan Sikk and Daniel Bochsler.

Note: Model weighted by the absolute number of participating voters in municipalities.

*Estimates based on 2000 census figures of Eastern Slavs and a constant citizenship rate of 50% among Eastern Slavs.

and ethnic data from the 2000 population census. Although the social structure of the municipalities may have changed since 2000 – especially as a result of the growth of commuter areas around Tallinn – many of the new arrivals are probably still registered as voters elsewhere.[20] As expected, we see a strong negative relationship between the percentage of Eastern Slavs among citizens and the Green vote, indicating an almost universally ethnic Estonian support base. The party also tends to be more popular in more affluent municipalities – the model predicts a difference of 5.9 percentage points between the most and least affluent municipality *ceteris paribus*. The candidacy of Marek Strandberg – the party's main initiator – in Harju and Rapla County added slightly less than a percentage point to the Green vote.

Figure 6 shows the geographic distribution of Green Party support. The pattern of support corresponds well with our analysis of survey and ecological data. Ida-Virumaa and Tallinn, where the proportion of Russian-speakers is higher, exhibit lower levels of support for the party – in the case of Tallinn, compared to surrounding areas. Also, in most rural counties with high numbers of people employed in agriculture, the support is lagging behind average. Three counties in northern and north-western Estonia have support levels clearly above average. That can be explained by a mixture of candidacy and structural effects. On the one hand, the commuter areas on the outskirts of Tallinn are home to many who enjoy clearly above-survival living standards. On the other hand, Harju and Rapla County formed the electoral constituency where both Strandberg and Valdur Lahtvee (the leader of the parliamentary group) contested the election. The high level of support enjoyed in Lääne County is evidently an effect of the popularity of Aleksei Lotman, a locally well-known biologist who worked for years in one of the main Estonian natural reserves there.

Above we attempted to discern the structural characteristics of Green voters. However, it is human beings not social structures that vote, and in order to get a better understanding of the Green Party we need to look into the interests, motivations and values of its supporters. Figure 7 shows the self-placement of Green

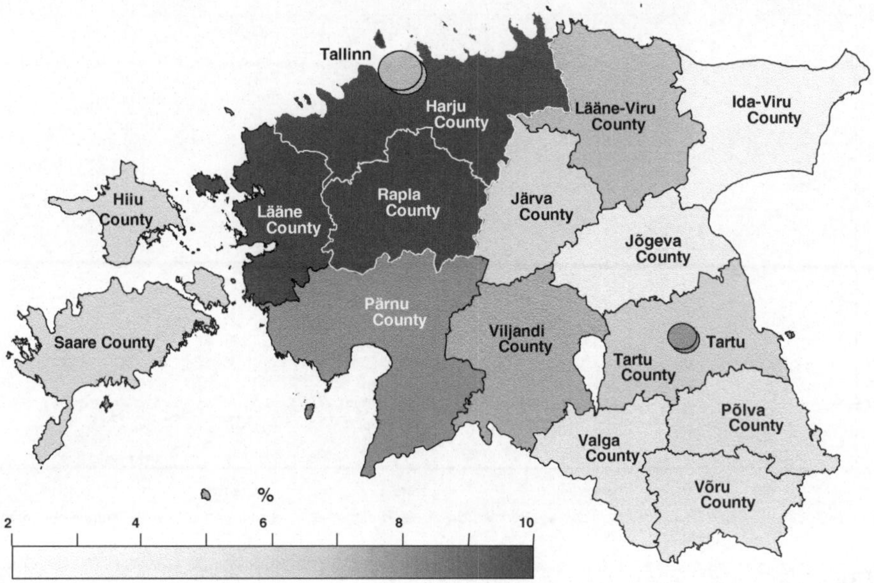

FIGURE 6 Geographic distribution of Green Party support. *Source*: Estonian National Electoral Commission (available at: www.vvk.ee, accessed 23 April 2009).

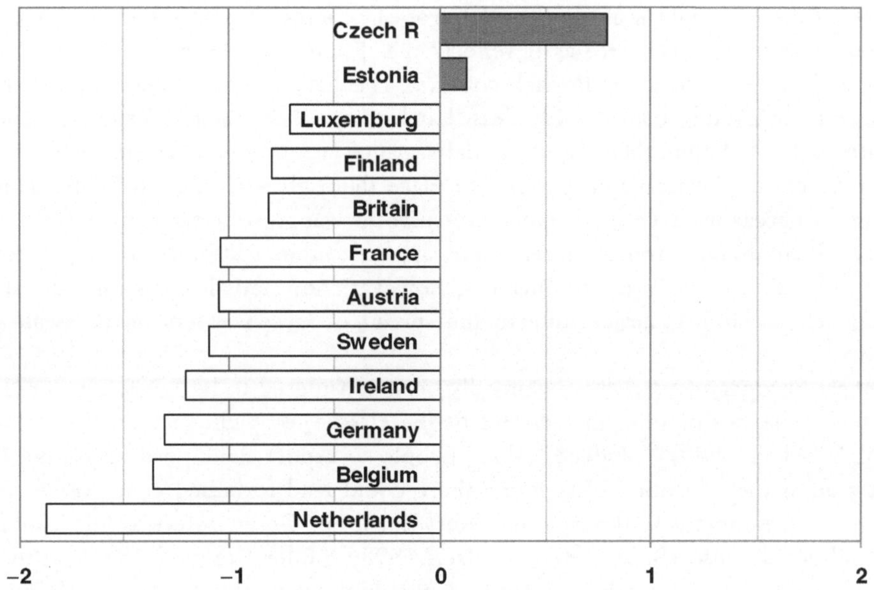

FIGURE 7 Self-placement of Green Party supporters on left–right scale against national average (10-point scale). *Source*: European Election Study 2004, Green Party supporters in European Elections. Estonian election survey 2007, Green Party supporters in national elections.

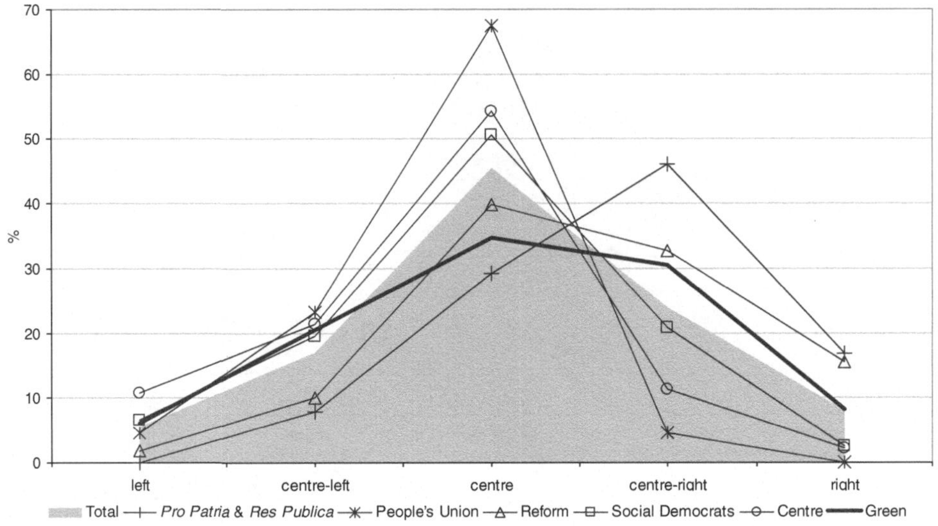

FIGURE 8 Self-placement of Estonian parties' voters on left–right scale (frequency distributions on a five-point scale). *Note*: the original 10-point scale has been transformed to a five-point scale for smoothing.

voters on the left–right scale in the 12 European countries with the strongest Green parties. The scale is relative, as it reflects the distance between the average Green voter and the average voter in each country. The Estonian and Czech Green voters stand out by being considerably less left-leaning than their Western European counterparts. Although there is substantial variance around the means in each country, Green voters in Western Europe tend to place themselves to the left of the national average, whereas the average Estonian Green voter was close to the centre of Estonian politics. If anything, Estonian Green voters appear to the right of the average Estonian voter. The difference from the mean is, however, not statistically significant at the 95% level ($p = 0.33$), largely due to the spread of Green voters on the scale (see below).[21]

West European Greens have generally been considered to be 'New Left' or 'Left-libertarian' parties or ones that do not fit the classic left–right scale well (Kaelberer 1993; Kitschelt 1988).[22] Indeed, asking people to report their position on the left–right scale if they do not feel at home there could lead to biased conclusions. Yet it does seem meaningful to describe the West European Green voters as left-leaning in the traditional sense (Birch 2009; Knutsen 1998; Müller-Rommel 1985). Although the relative positions of their voters on the left–right scales vary, the frequency distributions of their supporters clearly peak to the left of national averages. Placing 'the average' Estonian Green voter on the left–right scale is controversial because the deviation among the respondents is very high. Figure 8 shows that even though the distribution of Green voters on the ideological scale peaks slightly to the right of center, a considerable proportion of Green voters place themselves to the left of the middle position. This indicates that the classic left–right scale fails to define the

Estonian Green Party, as has also been noted by members of the party;[23] alternatively, the Green Party has a fairly heterogeneous support base.

According to the first model of new party emergence that was outlined earlier, the political system reflects the conflict lines that exist in a given society, and the development of new parties would normally be a result of changes in the underlying social cleavage structure. During normal times, the cleavage structures remain intact and so does the party system. In rare circumstances new cleavages emerge which can lead to the development of new parties if the existing parties fail to incorporate the new lines of conflict. The emergence of Green parties in Western Europe has often been explained along these lines. Echoing Inglehart's postmodernization thesis (1997), the proponents of the cleavage hypothesis explain the emergence of Green parties by the development of a new post-materialist cleavage (e.g. Kitschelt 1988). While people used to prioritize subsistence and material welfare, younger generations became increasingly concerned about non-materialist issues such as self-realization, political participation, human rights, gender issues and global problems. The development of post-materialist values was especially notable among the young and educated urban middle classes (e.g. Frankland 1995, p. 317). If the emergence of the Estonian Green Party is to be related to postmodernization, we should expect its voters to be policy-oriented in their reasons for voting, to express concerns for non-material values and be more interested and active in politics.

Alternatively, the emergence of the Green Party can be explained by a combination of a protest vote and political entrepreneurs searching for an opportunity to enter the political scene. There is some evidence of business interests behind the Green Party. Wind-turbine energy businesses were major donors to the party before the 2007 *Riigikogu* elections (Ilisson 2007).[24] In this view the new party does not result from new conflict lines but rather from a reaction against the status quo. Political stagnation, scandals, corruption and nepotism may prompt voters to look for new parties – not necessarily because they want policies to change, but because of dissatisfaction with the incumbents. New parties may win support by offering new faces without radically changing policy paradigms. If this were the case, the value orientation of Green voters should not differ from those of the supporters of other parties; nor do Green voters have to be politically or socially more active than the average voter.

The manifesto of the Green Party and public statements by its leaders send mixed messages about the issue of post-materialism beyond the concern for the environment. On the one hand, the promotion of democracy features fairly prominently in the Green manifesto. The need to increase the involvement of citizens and civil society organizations in policy-making is stressed ('Programmi põhiseisukohad' 2007, §6, §7). On the other hand, the Greens' positions on alcohol, tobacco, soft drugs and gay rights have been markedly conservative. The party favors the banning of alcohol and tobacco advertising, a substantial increase in excise taxes ('Rohelise Erakonna . . .' 2006, §7.2), and keeping the ban on soft drugs such as cannabis ('Vastused küsimustele' 2007). It is also opposed to gay marriage and adoption rights:

> Homosexual relationships will not be regulated by the state. It is a private matter and such relationship will not be considered a marriage and consequently no

rights for adoption will be granted to such co-habiting people. ('Rohelise Erakonna . . .' 2006, §1.3)

We do not have time-series data on the development of post-materialism in Estonia. As we saw in the previous section, Green voters tend to be ethnic Estonians, somewhat more educated and more urban than the average voter. Age is only weakly related to the propensity to vote Green. Some of these socio-economic characteristics may be interpreted as evidence in support of the post-modernist cleavage model, others contradict the general expectations. Higher levels of education and urbanization fit Inglehart's account of post-materialist individuals, whereas the absence of age effect runs counter to it.

Data from the 2007 Estonian Election Survey allow us to look at motivations driving the Green voters. Below we will analyze data on reasons for voting in the 2007 *Riigikogu* election, level of interest in politics, salience of different policy issues and engagement in various kinds of social and political activities.

Table 3 reports the main reasons for voting among the supporters of the Greens and other main parties. The battery of items included in the survey seemed to be sufficiently broad, as reasons not originally listed were mentioned by only 0.7% of the respondents. The wish to participate in the selection of the political leadership was the most frequently reported reason among supporters of other parties (28.5% of all respondents). The Green voters stand out here as only 15.3% reported that as the main reason, whereas the purpose of supporting a particular party was mentioned by nearly a third – three times more often than on average. Particular candidates and the tradition of voting were less important for the Green supporters than for those of other parties. None of the Greens included in the sample claimed that they had voted to support any particular candidate.[25]

When asked specifically about whether their vote was cast primarily for a particular candidate or a particular party or equally for a candidate and a party, 62% of the Green voters mentioned *mainly party* as their main motivation. No other party came close to these figures – among other parties, the 40% reported by the Reform Party supporters was the highest. Interestingly, a similar phenomenon occurred with *Res Publica*, the newcomer party of the 2003 election. The voters supporting *Res Publica* – which won

TABLE 3 Self-reported reasons for voting in 2007 election (%)

	Green	Other parliamentary parties
Wanted to support a particular party	30.5	10.2
Feel a part of my state	16.9	17.4
Want to participate in the selection of leaders	15.3	28.5
Voting is a citizen's obligation	11.9	12.1
A way to express my attitude	10.2	2.6
By voting, I can influence social processes	6.8	5.7
I have always voted	1.7	5.0
Wanted to support a particular candidate	0.0	11.9
Other reason	6.8	6.5

more than a quarter of votes – were also less prone to report a candidate as the decisive factor. However, *Res Publica* supporters were more concerned about selecting leaders, voting more out of habit and less focused on expressing their attitudes.

The Green voters stand out as being more concerned about expressing their attitudes through casting a ballot than is the average Estonian voter. This could be seen as connoting a willingness to take an active part in shaping Estonian political developments. One interpretation of this is that Green voters are more concerned with policies offered by the party than with who represents the party in the parliament.

Table 4 reports the respondents' opinions about important objectives of the Estonian society. Each respondent could choose the three most important ones from a list of 13 items. Unfortunately the 2007 Estonian Election Survey did not include items directly touching upon 'Greenness'. However, only very few respondents picked 'something else', indicating that the set list should cover the spectrum of important objectives rather well. Overall, the respondents mentioned 'economic security', 'job security and fair pay' and 'efficient and good health care' as the most important objectives. At least the first two of these items would normally be considered to reflect 'materialist' values. Green voters likewise mentioned 'economic security' and 'job security and fair pay' as the two most salient topics, though with somewhat lower frequencies. After these, three concerns were equally important for more than a quarter: 'preservation of national identity and culture', 'independence and security' and 'efficient and good health care'. However, the latter was mentioned much less frequently by Greens than most other parties' voters. Only the supporters of the *Pro Patria* and *Res Publica* Union expressed less concern for the health care system.

TABLE 4 Important objectives in Estonian society (%)

	Green	Other parliamentary parties
Economic security for oneself and relatives	48.3	53.1
Job security and fair pay	35.0	45.4
Efficient and good health care	26.7	35.8
Preservation of national identity and culture	26.7	23.7
Independence and security	26.7	21.7
Democratic rights and freedoms, participation in decision-making	25.0	11.6
Good education and retraining opportunities	23.3	29.1
National security; law and order	20.0	28.2
Opportunities for self-fulfilment and career	20.0	10.7
Limiting inequality of wealth	16.7	17.2
Strengthening of social solidarity	13.3	6.5
Integration of Estonia into the international community	8.3	5.3
Gender equality	1.7	2.5

Note: Each respondent could choose three answers from a set list.

The importance of topics such as 'preservation of national identity and culture' and 'independence and security' among the Green voters may come as a surprise. These are usually considered to be conservative values that run largely counter to the globalist and pacifist focus of most West European Green parties. However, taking the historical origins of Estonian Greens discussed earlier into account, the patriotic sentiments of its supporters are explicable. As Frankland (1995, p. 317) contends: 'in Western Europe, Greens coalesce in opposition to capitalist systems, while Greens in Eastern Europe emerged as opposition forces to communist systems'. The Estonian Greens are also different from most of their West European counterparts by being in favor of the country's NATO membership.[26]

At the same time, the Green voters attach more importance to questions such as 'democratic rights and freedoms, participation in decision making' (mentioned by 25%), 'opportunities for self-fulfilment and career' (20%), 'strengthening of social solidarity' (13.3%) and 'integration of Estonia into the international community' (8.3%). These objectives can be classified as post-materialist. It is, however, noteworthy that the Green voters do not consider gender equality to be an important issue. In fact, as another question in the survey indicated, more than 60% of its supporters disagreed with the statement that 'it would be beneficial to have an equal number of men and women in politics' – the highest proportion recorded among parliamentary parties. This rather striking result calls for further investigation.

We would normally expect post-materialist voters to be politically more active than voters with predominantly materialist views. Grass-roots activities and NGO membership have generally been considered important for post-materialists. If the Green voters are to be post-materialists, they ought to be both interested and active in politics. However, Figure 9 clearly shows that they do not stand out

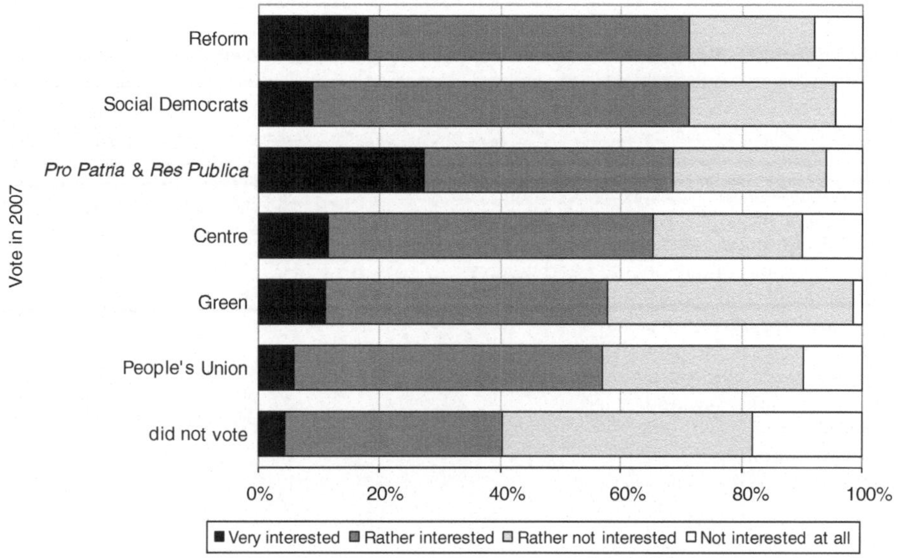

FIGURE 9 Self-reported interest in politics among party voters.

among the supporters of main parties and rather report below-average levels of political interest.

Although in its program the Green Party propagates extending the role of civil society in the political process, by most accounts their voters are politically not more active than other voters. In some respects, they are less enthusiastic – e.g. less prone to take part in solving local problems. There is, however, one area of activity where the Green voters visibly stand apart from the rest – they score significantly higher when asked whether they had purchased or boycotted goods or services for a cause during the last year. Even though these activities were exercised by less than half of the Green voters, there seems to be a group of 'conscientious consumers' among them.

Discussion and conclusion

It seems easier to establish what the Estonian Green Party is *not* than to determine what it *is*. This is partly a result of the newness of the party as a parliamentary actor.[27] We may be trying too hard to find evidence of post-materialism for the sheer reason that this was behind the advent of West European Green parties. Likening one to the other seems rather inappropriate. Even if environment is the main concern for the parties and they may belong to the same European federation, they appear to be so different deep down that placing them within the same party family is problematic. The common denominator of names disguises the fundamentally different logic underlying the development of the parties. While the West European Greens have generally been highly critical towards capitalism, for example advocating a radical redistribution of the world's wealth, the Estonian Green Party accepts capitalism and free enterprise and tends to focus on problems at the national rather than the global level.

Overall, the Green Party has features from *all* of the models of new party emergence outlined at the beginning of this study. Even though the environment was low on the political agenda during the 2004 European Parliament Elections when Marek Strandberg first voiced his intentions of re-establishing the Green Party, later developments – natural disasters and concerns over energy security – contributed to the growing importance of ecological concerns. Hence, the salience of the issue has increased, but it is difficult to see how an issue can convert into a real social cleavage if not accompanied by deep change in social structures or value orientations. The profile we could present of Green Party supporters does not show broad streaks of conventional post-materialism, but we should not assume a strict relationship between environmental concerns and post-material orientations. The weak presence of minority, gender and international development concerns in the manifesto of the Green Party and from the agenda of its voters is explicable when we take into account historical factors.

Different historical origins can easily explain why the Estonian Greens – in contrast to left-leaning West European Green parties – are located to the right of the political center. The socialist experience in Estonia was coupled with practices highly detrimental to the environment. The transition to a market economy, on the other hand, has brought along significant environmental improvements. The Soviet era was a period of mass industrial and agricultural production and army bases that paid scant attention to the well-being of the environment. In many places it left a long-lasting legacy and even though the memories of the Soviet era are perhaps slowly starting to wane – especially among the younger generations – the destroyed environment constitutes a lasting reminder of the adverse relationship between socialism and ecology. Interestingly, the transition to capitalism saw a weakening of the pressures on the environment. Heavy industries were either dismantled or subjected to new and much stricter pollution standards. There was also a general slump in agriculture accompanied by movement towards smaller-scale production and a decline in the use of mineral fertilizers. In many ways, the culprits for destruction of natural habitats have been very different in Estonia compared to Western Europe. Only more recently have the objectives of economic efficiency and the protection of environment started to clash more visibly. There is some evidence that an element of global or cross-border awareness has started to take off, indicated by the higher than average willingness of Green voters to purchase or boycott goods for a cause.

However, both the party program and the statements of its leaders make it evident that when it comes to furthering the ecological cause, the party prefers technological innovation over the post-materialist idea of constraining economic growth. In other words, rather than fighting the *modern* disease with *post-modernism*, the Estonian Greens try to take a less confrontational stance and believe that *ultra-modernism* offers a positive-sum solution that makes both consumers and the environment better off.

Acknowledgements

The authors wish to thank Daniel Bochsler, Seàn Hanley, an anonymous reviewer and the participants of the 2008 BASEES conference for constructive feedback on earlier drafts of the essay. An earlier version of the essay was presented at the BASEES Conference, 29–31 March 2008, Fitzwilliam College, Cambridge. The research for the paper was partly financed by the Estonian targeted finance project SF018012s08.

Notes

1 We do not intend to give a full account of pre-independence Green movement – its story has been well covered by Rein Taagepera (1993) and Matthew Auer (1998) among others. In Estonian, Juhan Aare (1999) presents a detailed insider's account of events from 1971 to 1989.

2 In fact the first battles of the 'Phosphate War' took place as early as 1971 among central authorities and some local leaders and Estonian scientists (Aare 1999).

3 Authors' interview with V. Lahtvee, Tallinn, 20 March 2008.

4 These outings have continued to the present day: in 2007 the eighteenth expedition took place.

5 The Bronze Soldier made headlines in the international press as it was relocated amidst rioting in April 2007.

6 Even though the stated aim of the Royalists was to create a constitutional monarchy and the party was known for its antics in the parliament, they were a protest party rather than a joke.

7 Authors' interview with V. Lahtvee, Tallinn, 20 March 2008.

8 1999 was the year when the Green Party disappeared, but also when the country recovered from the aftershocks of Asian and Russian economic crises.

9 Russia cut gas exports to Ukraine in the beginning of 2006. A few weeks later, explosions caused disruptions in the gas supply to Georgia (Hauser & Kernic 2006, pp. 155, 158).

10 *Postimees*, 21 July 2007.

11 Authors' interview with V. Lahtvee, Tallinn, 20 March 2008. Estonia has met its electricity needs from oil-shale-based power plants in the north-east of the country. However, that comes at the cost of substantial CO_2 emissions and oil shale will lose its competitive edge once electricity markets are opened up according to EU rules. For more, see a report by Estonian Foreign Policy Institute (2006, pp. 11, 16).

12 Regularly updated lists of party members are publicly available over the internet (https://ar.eer.ee/erakonnad.py, accessed 23 April 2009). In order to be secure against possible mass defections the 'sustainable' number of party members is even higher.

13 Ten years earlier he had been acquitted in a high-profile court case related to the sale of roubles following the 1992 currency reform. The original scandal led to the resignation of Prime Minister Mart Laar in 1994 (*Estonian Court Acquits . . .* 1996).

14 Authors' interview with V. Lahtvee, Tallinn, 20 March 2008.

15 This strategy has been advocated for studying Greens and other small parties by Birch (2009).

16 In 2008, the statute of the party was changed to establish a single leader.

17 Authors' interview with V. Lahtvee, Tallinn, 20 March 2008.

18 Both the Reform Party and the *Pro Patria* and *Res Publica* Union are clearly on the right of center. Even though the Social Democrats have moved left in recent years, it is still a relatively rightist social democratic party by European standards.

19 The share of non-Estonians among the population of the five largest cities is considerably higher than among the population at large.

20 The electoral rolls in Estonia are based on constantly updated data from population registers. Changing registration after moving from one municipality to another is not strictly compulsory and is often not done out of neglect or because of benefits that go with it (i.e. childcare and schools).

21 The same applies for the Czech Republic ($p = 0.12$).

22 There are minor exceptions of environmental parties that are not clearly on the left – e.g. the environmental niche in Sweden has been occupied by the Centre Party (Rohrschneider 1993).

23 For example Tammert (2007) and the authors' interview with V. Lahtvee, Tallinn, 20 March 2008.

24 Nevertheless, the electoral campaign of the Green Party paled against those of other parties entering the parliament (*Kokkuvõte. . .* n.d.).

25 Elsewhere in the essay we have discussed the impact of particular candidates on Green Party support. That does not contradict our findings here. On the one hand, a substantial number of Green voters still gave some importance to

particular candidates. On the other hand, the Estonian ballot structure makes voting for a party without prominent candidates in the district tricky. Even though the candidates are organized by party lists, the voters can only cast their vote for a particular candidate. Hence, if a voter does not recognize any of the candidates, the incentive for voting for the party may decrease.

26 'Programmi põhiseisukohad', §43, available at: http://wiki.erakond.ee/images/4/45/EERuusprogramm.pdf, accessed 24 April 2009.

27 We have a reason to be optimistic about the chances of studying further development of the Green Party – its support endured or even increased in public opinion polls after the elections, reaching 12% in early 2008 (EMOR 2008).

References

Aare, J. (1999) *Fosforiidisõda 1971–1989* (Tallinn, Kirilille Kirjastus).

Africa Confidential (2006) 'Côte d'Ivoire: The toxic trade', 47, 20, pp. 4–5.

Alvela, A. (2005) 'Looduskaitsjad väldivad kohalikku toetusraha', *Äripäev*, available at: http://ap3.ee/Default2.aspx?PaperArticle=1&code=3041/new_eri_artiklid_304115, accessed 26 March 2008.

Antsov, A., Kool, A. & Talviste, V. (1998) 'Erakond Eesti Rohelised, saksa rohelised ja muu maailm', Department of Political Science, University of Tartu, unpublished manuscript.

Auer, M. R. (1998) 'Environmentalism and Estonia's independence movement', *Nationalities Papers*, 26, 4, pp. 659–76.

Birch, S. (2009) 'Real Progress: Prospects for Green Party Support in Britain', *Parliamentary Affairs*, 62, 1, pp. 53–71.

Beyme, K., von (1985) *Political Parties in Western Democracies* (Aldershot, Gower).

EBRD (2008) *Economic statistics and forecasts*, available at: http://www.ebrd.com/country/sector/econo/stats/index.htm, accessed 28 March 2008.

EMOR (2008) *Toetused erakondadele*. available at: http://www.emor.ee/failid/erakonnad.html, accessed 26 March 2008.

ENS News (2006) 'Three Baltic States Say "Yes"', No. 12, available at: http://www.euronuclear.org/e-news/e-news-12/baltic-states.htm, accessed 27 March 2008.

'Estonian Court Acquits Defendants of 1992 Ruble Sale' (1996) *F&P RFE/RL Archive*, 24 June.

Estonian Foreign Policy Institute (2006) *Energy Security of Estonia in the Context of the Energy Policy of the European Union*, Research paper commissioned by the *Riigikogu* Foreign Affairs Committee, available at: http://www.evi.ee/lib/Security.pdf, accessed 26 March 2008.

Frankland, E. G. (1995) 'Green Revolutions? the Role of Green Parties in Eastern Europe's Transition, 1989–1994', *East European Quarterly*, 29, 3, pp. 315–45.

Geyer, G. A. (1988) 'Baltic Reforms New Puzzle for Russians', *The Oregonian*, 9 November.

Hauser, G. & Kernic, F. (2006) *European Security in Transition* (Aldershot, Ashgate).

Hopkin, J. & Paolucci, C. (1999) 'The Business Firm Model of Party Organisation: Cases from Spain and Italy', *European Journal of Political Research*, 35, 3, pp. 307–39.

Ilisson, A. (2007) 'Rohelisi toidavad tuuleenergeetikud', *Äripäev*, available at: http://ap3.ee/Default2.aspx?PaperArticle=1&code=3454/uud_uudidx_345401, accessed 27 March 2008.

Inglehart, R. (1977) *The Silent Revolution: Changing Values and Political Styles Among Western Publics* (Princeton, Princeton University Press).

Inglehart, R. (1997) *Modernization and Postmodernization: Cultural, Economic, and Political Change in 43 Societies* (Princeton, NJ, Princeton University Press).

Järlik, R. (2004) 'Balti vabaduse tee', *Maaleht*, available at: http://www.maaleht.ee/?page=Arvuti&grupp=artikkel&artikkel=1686, accessed 26 March 2008.

Kaelberer, M. (1993) 'The Emergence of Green Parties in Western Europe', *Comparative Politics*, 25, 2, pp. 229–43.

Karm, E. (2008) 'Environment and Energy: The Baltic Sea Gas Pipeline', *Journal of Baltic Studies*, 39, 2, pp. 99–121.

Kärmas, M. (2001) 'Preestrist eriuurija', *Eesti Ekspress*, 15 February.

Kitschelt, H. P. (1988) 'Left-Libertarian Parties: Explaining Innovation in Competitive Party Systems', *World Politics*, 40, 2, pp. 194–234.

Knutsen, O. (1998) 'Expert Judgements of the Left-Right Location of Political Parties: A Comparative Longitudinal Study', *West European Politics*, 21, 2, pp. 63–94.

Koch, T. (2007) 'Riigikogu valimised: Roheliste erakond kasvatas vahespurdiga oma toetust', *Postimees*, 6 February.

Lipset, S. & Rokkan, S. (1967) 'Cleavage Structures, Party Systems, and Voter Alignments: An Introduction', in Lipset, S. & Rokkan, S. (eds) (1967) *Party Systems and Voter Alignments: Cross-National Perspectives* (New York, Free Press), pp. 1–64.

Lucardie, P. (2000) 'Prophets, Purifiers and Prolocutors: Towards a Theory on the Emergence of New Parties', *Party Politics*, 6, 2, pp. 175–85.

Mair, P. (2007) 'Left-Right Orientations', in Dalton, R. J. & Klingemann, H.-D. (eds) (2007) *The Oxford Handbook of Political Behaviour* (Oxford, Oxford University Press), pp. 206–22.

Müller-Rommel, F. (1985) 'The Greens in Western Europe: Similar But Different', *International Political Science Review*, 6, 4, pp. 483–98.

Müller-Rommel, F. (ed.) (1989) *New Politics in Western Europe: The Rise and Success of Green Parties and Alternative Lists* (Boulder, CO, Westview Press).

Poom, R. (2007) 'Ignalina tuumaprojektis osalemine ei pruugi Eestile olla parim valik', *Eesti Päevaleht*, 6 August.

Rochon, T. R. (1985) 'Mobilizers and Challengers: Toward a Theory of New Party Success', *International Political Science Review*, 6, 4, pp. 419–39.

'Rohelise Erakonna' (2006) 'Rohelise Erakonna Algatusgrupi programmilised plaanid', 17 June, available at: http://wiki.erakond.ee/index.php/Rohelise_Erakonna_Algatusgrupi_programmilised_plaanid, accessed 24 April 2009.

'Rohelised soovivad energiapoliitika muutmiseks erakorralist istungit' (2007) *Postimees*, 12 July.

Rohrschneider, R. (1993) 'New Party versus Old Left Realignments: Environmental Attitudes, Party Policies, and Partisan Affiliations in Four West European Countries', *The Journal of Politics*, 55, 3, pp. 682–701.

Rudi, E., Mägi, M., Link, E.-G. & Maimets, A. (2000) 'Eesti roheliste kadumine poliitiliselt maastikult', Department of Political Science, University of Tartu, unpublished manuscript, available at: http://www.ut.ee/ABVKeskus/muu/Roheliste.doc, accessed 26 March 2008.

Select Committee of Riigikogu on the Application of Anti-Corruption Act (n.d.) *Kokkuvõte 2007. aasta Riigikogu valimistel osalenud erakondade ja üksikkandidaatide valimiskampaania rahastamise aruannetest*, available at: http://www.riigikogu.ee/public/Riigikogu/Korruptsioon/kokkuv_te_RKval2007.pdf, accessed 27 March 2008.

Sikk, A. (2006) *Highways to Power: New Party Success in Three Young Democracies* (Tartu, Tartu University Press).

Sikk, A. (2007) 'Good to be "New": On Advantages and Disadvantages of New Political Parties', Working paper presented at the workshop 'Populism in Enlarged Europe' 6–7 December, Stockholm, Södertörns Högskola.

Sikk, R. (2006) 'Andrus Ansip, sõitke Tšernobõli', *Eesti Päevaleht*, 26 April.

Smith, M. (2006) 'Czech Republic: Greens Take Root', *Transitions Online*, 05, 23.

Soomere, T. & Quak, E. (2007) 'On the potential of reducing coastal pollution by a proper choice of the fairway', *Journal of Coastal Research*, SI 50, pp. 678–82.

Taagepera, R. (1993) *Estonia: return to independence* (Boulder, Westview Press).

Taagepera, R. (2006) 'Meteoric trajectory: The Res Publica Party in Estonia', *Democratization*, 13, 1, pp. 78–94.

Tammert, P. (2007) 'Rohelised vasak-parempoolsel skaalal', *Postimees*, 23 March.

Toomla, R. (1999) *Eesti Erakonnad* (Tallinn, Eesti Entsüklopeediakirjastus).

'Vastused küsimustele' (2007) 'Vastused küsimustele', available at: http://wiki.erakond.ee/index.php/Vastused_küsimustele, accessed 23 April 2009.

MINORITY ENVIRONMENTALISM AND ECO-NATIONALISM IN THE BALTICS: GREEN CITIZENSHIP IN THE MAKING?

Tove H. Malloy

Introduction

The discourse of environmental protection is increasingly being enlisted as part of minority politics. This was certainly evident in the Baltic republics during the last years of the Soviet regime; another case in point is the movement of the Zapatistas in Mexico, which carries a similarly strong undertone of sovereignty and self-determination. Yet less radical national minorities also evidence desires to protect the Earth. Creating bioenergy, environmentally managing biodiversity and homelands, as well as creating grass-roots movements following the 'think globally, act locally' mantra are just a few examples of the efforts that mostly territorially defined minorities[1] have embraced for various reasons and ends. Yet history also shows that in cases such as the Baltic states, where sovereignty and self-determination were in focus, the environmental approach was heavily informed by nationalistic (Schwartz 2006, p. 201) and rather opportunistic special-interest politics (Galbreath 2005; Hosking 1992) once independence was won. Minorities who became rulers forgot about protecting the Earth in favor of protecting their newly gained or regained nation (Dawson 1996; Galbreath & Auers 2009). In short, nationalistic politics pushed environmentalism on to the back-burner.

If nationalism could emerge so smoothly out of the environmental discourse, one must ask whether this was motivated by a natural connection between environmentalism and nationalism, or if it was, as the facts seem to suggest, simply a calculated or reframed freedom strategy (Dawson 1996). Either way it was a matter of identities that became motivated enough for collective action and ultimately for ideological reasoning. Of course, circumstances and contingencies as well as historical legacies determine much of what happens in such situations. The environmental politics pursued by the titular peoples[2] in the Baltic states during the final days of the Soviet era was highly complex and involved the entire gamut of imperial and independence politics. But why was environmental discourse the mobilizing force? Are there any logical links between environmentalism and nationalism or environmentalism and freedom? This essay will discuss these questions by querying Green political theory's relationship to both nationalism and justice.

It is no secret that this relationship has been an uneasy one. The narratives of Green political thought contain very few words on nationalism, while justice seems to constitute an eternal conundrum (Malloy 2007). Nationalism is seen first and foremost as too instrumental for genuine environmentalism, and justice in the sense of self-determination over territory as well as boundaries that delineate sovereign territory is inimical to the globalizing of environmentalism (de-Shalit 2006). Rather, sovereignty must give way in order to optimize environmental protection and make it globally sustainable (Deudney 2006; Hurrell 2006; Linklater 2006). Moreover, freedom and rights (Dagger 2006) as well as social justice (Dobson 1998) are seen as legacies of liberalism that must be accommodated by Green theory *in so far that they ensure sustainability* (Bell 2002; Dobson 1990, 2003; Wissenberg 2006). This would seem to indicate that there is room for accommodating freedom struggles within Green theory as long as they do not enlist the nationalist discourse. Nationalism pursues sovereignty while environmentalism rejects it. It would therefore appear that a transfer from an environmental strategy to a nationalist one in pursuit of sovereignty is not feasible within Green theory because it jeopardizes sustainability.

But sovereignty by way of environmentalism is precisely what the titular peoples of the Baltic states sought. I therefore examine environmental justice and nationalism to assess how an environmental strategy can be informed by desires for national self-determination and sovereignty. Having examined the theoretical issues, I turn to minority environmentalism in practice, discussing briefly a few cases outside the Baltic states before examining developments in environmental politics in the Baltic region during the last days of Soviet domination. This discussion will lead to a consideration of how a co-optation of liberal justice by the environmental discourse in the Baltic states may in fact have formed a bridge to eco-nationalism.

What is environmentalism?

Green political theorists[3] operate with two types of Green thinking – ecologism and environmentalism. While environmentalism argues for a managerial approach to environmental problems, believing that these can be solved without fundamental changes in present values or patterns of production and consumption, ecologism holds that a sustainable and fulfilling existence presupposes radical changes in our relationship with the non-human natural world and in our mode of social and political life (Dobson 1990, p. 2). Environmentalism is not an ideology because it does not provide an analytical description of society, it does not prescribe a particular form of societal beliefs, and it does not provide a program for political action. As such, environmentalism may be accommodated by other ideologies and does not necessarily become a strand of ecologism. This seems to indicate that a relationship between environmentalism and nationalism might be feasible. Moreover, environmentalism represents a managerial approach to the environment within the context of present political and economic practices (Dobson 1990, p. 34). It does not necessarily subscribe to a limit to growth nor does it seek to dismantle industrialism. It does not argue for the intrinsic value of the non-human environment and rejects any attempt to reconstruct the human race metaphysically. It believes that technology can solve the problems it creates.

Ecologism, on the other hand, possesses a number of definitional tenets of ideology and is capable of being described as an ideal type. It purports to stand apart from other ideologies, such as liberalism (Wissenburg 2006) or socialism, and should not be seen as a crosscutting ideology taking aspects of tenets from other ideologies to create an eclectic, new fad of ideas.[4] The relationship of ecologism to political theory has been likened to that of feminism, and should therefore be seen as a 'radical Green challenge to the political, economic and social consensus that dominates contemporary life' (Dobson 1990, p. 2). In other words, where environmentalism seeks reform through a cleaner service economy sustained by cleaner technology and producing cleaner affluence, ecologism is based on a 'self-consciously hard-headed assessment of the unsustainability of present political and economic practices' (Dobson 1990, p. 11). Thus, normally environmentalism and ecologism need to be kept apart because they differ in degree and kind.

It could be argued, however, that in situations where minorities feel disenfranchised on all fronts, including the management of the natural environment of their homeland, the rational management approach of environmentalism might be informed by ecological tenets of society. Since environmentalism is not a theory of society, it may be instructive to bring in ecologism and analyze nationalism and justice as theorized in this body of narratives. As regards the Baltic cases specifically, it is therefore necessary to provide a more in-depth overview of ecologism and its relation to environmental justice and eco-nationalism.

'. . . in theory' – a Green ideology

The Green theory of ecologism is today a comprehensive body of political thought that proposes to deal with the environmental issues facing our planet in a different manner than most conventional theories of political order. Philosophically, Green theory is the only political theory that has a view about the relationship between human beings and the non-human natural world. Basically, Green theory holds that the non-human natural world is ethically important and should be defended intellectually. It is not just that the non-human world constitutes a set of resources for human use but that we should not treat the non-human world in a purely instrumental fashion. Green theory thus includes *a theory of value*, which holds that the natural value of things created by natural processes, rather than by artificial human processes, should be preserved and promoted (Goodin 1992, p. 27). Things refer here not only to species and the natural world but also processors of natural value. The intrinsic value of the natural is thus a core tenet of what might be called the moral philosophy of Green theory. By the same token, natural things must not be split or separated in the overall goal to preserve them. Or to put it another way, natural things must be seen as holistic in their being part of our ecosystem. Natural value is thus also the essence of the philosophical outlook that Green theory defends through the *theory of environmental sustainability*. The sustainable society principle is that the Earth is finite and implies scarcity, thus placing limits on industrial growth. A fundamental framework for a sustainable society must therefore take into account not only limits to growth but also aspects of consumption, energy, trade and travel, work patterns, regionalism, and agriculture, as well as decentralization and community.

Limits-of-growth politics follow three principles. First, technological solutions cannot help realize the impossible dream of infinite growth in a finite system. Second, the exponential nature of growth underpins its unsustainability, and third, the immense complexity of the global system has resulted in clumsy and superficial solutions so far (Dobson 1990, p. 68). The logical consequence for Green political theorists is to question our level of consumption in terms of need, population and technology. Moreover, they question our relation to energy in terms of both quantities of consumption and unethical application. Non-renewable energy is not considered healthy for a sustainable society. As self-reliance thus becomes an important tenet in the aim to keep our societies sustainable, Green political theorists advocate restraint on trade and travel. Trade is seen as wasteful, as creating vulnerable communities, as the locus for exercise of political and economic power through dependency and as creating unnecessary needs (Dobson 1990, p. 90). Finally, Green theory calls for a reconceptualization of the nature and value of work. Green political theorists believe in the value of work but question patterns of work in terms of early retirement (more people taxing the ecosystem); automated production (restraint on resources will bring back labor-intensive production); paid employment (many tasks that are not considered work and not paid for could be seen as work); and guaranteed basic income schemes (declining productivity decreases tax revenues which in turn results in no funds for guaranteed income schemes).

However, the future need not look bleak according to Green political theorists. The reason for this is that the problems caused by centralized systems of production and social security can be solved at the local level. *Bioregionalism* is thus a core tenet of Green theory. Bioregionalism refers to the idea that we get 'to know the land around us, learn its lore and its potential, and live with it and not against it. We must see that living with the land means living in, and according to the ways and rhythms of, its natural regions – its bioregions' (Sale 1985, p. 56). Bioregionalism includes various types of regions. Ecoregions are the largest type, usually several hundred thousand square miles; georegions are smaller, perhaps a few tens of thousands of square miles; and morphoregions or vitaregions are of several thousand square miles (Sale 1984, p. 227; 1985, p. 58). Bioregionalism involves identifying bioregional boundaries and living with what those territories provide in the way of natural resources and natural products. In bioregions people live in communities, which seek to minimize resource-use, emphasize conservation and recycling and avoid pollution and waste. Land is often communally owned and centralized institutions are avoided. In short, the guiding principle of bioregionalism is that the natural world determines the political, economic and social life of communities and that it reduces the spiritual and material distance between us and the land. While not all Green political theorists will subscribe uncritically to this picture, many will urge people to live in place, and to accommodate the environment rather than resisting it (Dobson 1990, p. 101). Agriculture therefore has a central place in bioregionalism and in Green theory.

Sound agriculture, according to Green theory, is not only non-polluting and organic; it is also labor-intensive and spiritual. Some Green political theorists prefer to promote decentralization rather than bioregionalism even though their arguments rest on much the same premises as those promoting bioregionalism (Bahro 1986; Goldsmith 1972). Both emphasize the advantages of small communities in terms of human relationships, more economical solutions, and more democratic decision-making. The *community* is thus the core element of the Green political society because it is within communities that attitudes to the need for environmental protection and social change can best be fostered. Finally, bioregionalism based on Green theory must ideally create societies that thrive on the diversity of human behavior, fostering democracy, freedom, tolerance and equality. Equality is seen as a holistic view of the science of ecology, which defends equality of status among species. A political ecology outlook would thus emphasize the inter-relationship and the inter-dependence between species and nature on the basis of an egalitarian outlook.

The upshot of this political outlook is that Green theory is profoundly anti-anthropocentric, meaning that the Enlightenment view that man is the center of the world is not a presentable picture of reality. *Anthropocentrism* as a world view is 'held to be a basic cause of environmental degradation and potential disaster' whether in its weak meaning of human-centered or its strong meaning of human-instrumental (Dobson 1990, p. 51). The non-anthropocentric view of ethics thus requires followers of Green theory to conform to a set of ethics based on the one hand on a code of conduct, and on the other on a state of being. The ecological *code of conduct* is informed by what has been termed 'deep ecology' or the concern for ecological principles such as complexity, diversity and symbiosis (Naess 1973, p. 95). This means the belief that the non-human world could have and does have intrinsic value, or an attempt to move beyond human-prudential arguments for concern for the biosphere (Dobson 1990, p. 46). This idea has been the object of some controversy among Green political theorists as to scope and membership and the fact that it may result in an authoritarian approach to the theory of value. Generally, however, it is held that 'a change in the attribute that invokes moral considerability inevitably brings with it a shift in the boundaries of ethical concern' (Dobson 1990, p. 41). The theory of the *state of being* in Green theory is instead defined by an ethic that takes into consideration ecological consciousness, or the idea that we identify with the non-human world, that we presage our self-realization on such a view and that our behavior is a logical result from this view. In other words, it is an alternative view of reality based on a social critique rather than on a metaphysical view of ethics.

The social critique therefore yields a picture of the world that shows 'that present social relations and the goals and desires that spring from them are unsatisfactory, and that new conceptions of self-fulfillment and happiness are desirable'

(Thompson 1983, p. 98). This is why Green theory in terms of ecologism has positioned itself primarily as an alternative to the prevailing views of both local and global approaches to environmental management (Dobson 1990, p. 163). The Green *strategy for social change* is consequently aimed primarily at practices rather than institutions. As the strategy calls for a fundamental shift this renders Green theory rather more radical than the environmentalism described earlier. Since Green theory aligns itself with a democratic outlook, the strategy involves both parliamentary and extra-parliamentary action (Dobson 1990, p. 113). Parliamentary action is, however, fraught with problems as parliamentary survival often sets the agenda of political parties rather than the issues at stake (Bahro 1986, p. 211). The desire to change practices (of parliamentarians) has thus not been an easy goal for most Green parties. Instead, Green theory proposes to change the practices of lifestyles, communities, direct action and classes (Dobson 1990, p. 130). The strategy for social change is thus an implementation scheme of the rules of ethics. In terms of lifestyle, social change should be sought in the patterns of individual behavior in daily life. Typical examples of this are taking better care of, and with, the things that we buy; being careful in the things we say; paying careful attention to where we invest our money; and the way we treat other people. It can also include care about the means of transportation we use. In other words, Green consumerism is a core tenet in a process of social change. However, it has been argued that Green consumerism alone is simply another type of environmentalism, if it is not accompanied by a fundamental change of psyche that brings about a calmer, gentler, and 'Greener' state of consciousness (Dobson 1990, p. 133). Change on the basis of modified interests is therefore not enough. It must take the spiritual dimension of Green theory seriously.

For some Green political theorists this change is best promoted within the local communities referred to earlier. Revolutionizing community practices can support the aim of changing lifestyle practices. Some Green political theorists argue that community living is by far the most ideal mode by which to seek changes in social behavior. Communitarian behavior supports the inter-relational and inter-connectedness that the ecological principle of equality requires in order to achieve parity between the human world and the natural world. People living in communities are thus more likely to see the merits of seeking sustainability through improved social behavior, such as self-help, community responsibility and free activity that are consistent with ecotopian ideals of loose federations of regions and communes (Eckersley 1987, p. 22). On this view, community members see jointly the merits of the Green future that individuals in urban societies may not. This is not to argue that urban individuals are incapable of seeing the merits of a Green future. Certainly, social movements with environmental agendas do exhibit the morality of the urban individual becoming involved. However, direct action implemented by such movements often ends with the movement on the losing side of the battle. Instead, some Green political theorists have suggested that change may best happen as a class phenomenon. It may be that certain classes are more open to changing their social behavior towards the environment and Mother Nature in general. In a similar vein, it has been suggested that women could be the promoters of Green social change as they occupy a crucial space in the reproductive process and an optimal way to sense nature and respect nature is through your body (Salleh 1997, p. 38). Whichever strategy is followed it is clear that to Green political theorists, Green theory is a transformative political ideology (Salleh 1997, p. 111). It stands out by virtue of its foundational approach to the relationship between human beings and their natural environment, its belief in the limits to material growth and its non-anthropocentrism. In short, Green political theorists hold that Green theory offers a coherent critique of contemporary society and a prescription for improvement.

Environmental justice

As a prescription for improvement, Green theory focuses mostly on protection; protection of both species and ecosystems against current and future destruction. This is why Green political theorists want to temper a rights-based approach, insisting on a definitive view of the proper moral relationship between human beings and the non-human natural world. Green theory acknowledges the utility of rights-talk if it is employed for environmental ends and complemented with a notion of responsibilities (Dobson 1990, p 172). It asks not what we can get out of the world, but what we can do for the world (Hayward 1994, p. 163). As such, Green theorists would wish to sever the orthodox relationship between rights and duties by emphasizing responsibilities. Moreover, a Green conception of responsibility would require that this be seen in terms of individual autonomy. Green autonomy is thus self-government rather than license (Dagger 2006, p. 213). The Green conception of autonomy sees autonomous species as interdependent and receptive to the Green ethics of code of conduct and the state of being. According to this notion, people need not surrender their freedom in order to respond effectively to the ecological challenge. Although they must surrender some of their options in life to be good Green citizens, and thus some rights could be curtailed, they would apply a mode of practical reasoning about the environment in relation to individual autonomy and freedom which in turn renders the result ethical. Green justice is thus as much about ethics as it is about entitlements.

Green justice should not, however, be confused with environmental justice. According to Green theorists, environmental justice is a liberal-rights based approach to justice for those who are in a situation of environmental risk (Dobson 1998, p. 19). Poor people and those living in rural and remote areas are considered the primary beneficiaries of environmental justice, since they often live in places that are neglected or deliberately left to environmental degradation or used for extraction that degrades the environment. Environmental justice thus aims primarily to protect human beings and the protection of the environment is but an instrumental tool in this endeavor. Environmental justice furthermore claims liberal equality, which also puts it at odds with Green theory. In short, environmental justice is not in the business of achieving sustainability of the environment but rather sustainability of human beings.

Nationalism

Nationalism speaks both positively and negatively to Green theory. It is positive in terms of being an ideology that may foster solidarity and responsibility towards fellow citizens, and promote preservation of national heritages and a sense of obligation to future generations. On the negative side, nationalism may contradict thinking globally while acting locally; its mystical, abstract and instrumental attachment to territory and homeland may get in the way, thus preventing the cosmopolitan streak of Green theory from taking hold; it may not be able to transcend the idea of borders being ecological rather than political and historical; and it is anthropocentric (de-Shalit 2006). Avner de-Shalit's overall evaluation of the relationship between Green theory and nationalism is thus negative. While the positive aspects of nationalism make a case for coupling ecologism and nationalism, the negative aspects outweigh these. First, the 'act locally, while thinking globally' mantra is not promoted by nationalism, which puts emphasis on national interests rather than global interests. This, argues de-Shalit, is problematic especially when international environmental treaties are to be negotiated (2006, p. 82). Second, the mystical, abstract and instrumental attachment to territory and homeland, while it might encourage ruralism and bioregionalism, is a romantic glorification of rural life as superior to urban life that does not necessarily advance environmentalism (de-Shalit 2006, p. 82). Third, the political versus ecological definition of borders is problematic because it reifies the notion of sovereignty as the protector of a state's autonomy. By seeing sovereignty as over-riding the need to protect the environment as well as the need to make members of specific societies change their (bad) environmental behavior, sovereignty blocks the efforts of ecologists to address environmental problems on a global scale (de-Shalit 2006, p. 86). Environmental disasters do not respect political boundaries. Whether one agrees with de-Shalit or not, the problem of sovereignty is certainly a central aspect of the tension between Green theory and nationalism.

Green theorists see sovereignty as an Enlightenment legacy that is often driven by nationalistic tendencies. Sovereignty basically stands in the way of improvement of environmental protection and ecological politics. According to Green theorists, 'state sovereignty needs to be understood not as a discrete set of claims based solely on state power, but rather as a historically constituted "bundle of competences" whose character depends on the changing constitution of the international legal, political and economic order as a whole' (Hurrell 2006, p. 167). This 'Greening' of sovereignty is considered an enormously difficult process as the sovereign state continues to hold a dominant role in the facilitation of environmental change (Hurrell 2006, pp. 180–1). This is why sovereignty in relation to Green theory is important to our analysis of the transition in the Baltic states from the environmental justice discourse to the nationalistic discourse.

Finally, the fact that nationalism is rather anthropocentric means that it is unable to care for animal and non-human species on ethical terms. Nationalism, argues de-Shalit, treats non-human species as irrelevant unless they contribute to the nation's ethos by being symbols, such as trees or birds (2006, p. 88). In other words, nationalism would have little currency for Green state construction except in the local communities, where it could support an ethos of solidarity and responsibility towards both heritage and future generations.

'. . . in practice' – identities and ideology

It is precisely in the local community that minority environmentalism is observed. To the Zapatistas of Mexico it is the biodiversity of the Chiapas jungle that provides the identity for much of their independence struggle and thus their identity as indigenous people living amongst this biodiversity. The destruction of the jungle for oil extraction and large-scale logging were some of the core issues that motivated the Zapatistas. The leader of the Zapatistas, Subcomandante Marcos, has condemned practices of expropriation from the earth that are unhinged from any sense of indigenous reality (Becker 2006). More importantly, the uprising was also a reaction to the post-NAFTA flooding of the Mexican agricultural market with genetically modified corn. Since the Zapatistas regard corn as the original ancestor of all humans, its continued purity has profound importance for the preservation of their indigenous identity. Zapatista fighters were killed while guarding seed safe-houses where their heirloom seeds were preserved (Becker 2006, p. 17).

The close relationship of indigenous peoples to nature and the Earth is of course vividly illustrated by the plight of the Native American nations, which regarded expropriation of their territories as an encroachment not only upon their means of living but also upon their identity as one with their land. Native Americans' perception of the natural world and, thus, their approach to the management of biodiversity differs starkly from the Western ethnocentric and instrumental view (Stransky 2006). In Alaska environmental disasters are exacerbated because traditional ways have been pushed aside in favor of current practice. The miscommunication and lack of respect that arises from cultural misunderstanding adversely affect all aspects of disaster-planning, mitigation, preparedness, response and recovery. Stransky argues that native communities in Alaska have reverted to cultural traditions in order to deal with environmental emergencies, abandoning institutional (White-dominated) practices because of their inherent inefficiencies to date.

Correct environmental management of the homeland region is also the aim in South Tyrol, the northern-most region of Italy in which 75% of the population are German-speaking. The region has autonomy within the Italian state, and 90% of its tax revenues are returned by central government. This enables the government of the ruling party, the South Tyrolean People's Party, to enact legislation in support of preserving the Tyrolean identity and homeland. The industry that is the most taxing on the local environment is the tourist industry, as the region offers both summer and winter sports. Thus, the electricity that keeps the resorts open comes from water turbines, which represent a CO_2-neutral means of energy production (Lønborg 2009). Hotels are bioheated using only local biomass, which is also CO_2-neutral because it eliminates the need for oil trucks crisscrossing the small mountain roads. Burning biomass does have some waste consequences, therefore environmentally correct filters are fitted. In short, nature is the most important capital that the Tyrolean German minority has, and for this reason they have become environmentally conscientious. But more importantly, apparently the ownership of these resorts is kept in the minority's hands, thereby securing the 'sovereignty' of the autonomous region.

Homeland protection is precisely what explains the Green action, which has become part of the life of established national minorities in Europe. Like the indigenous peoples, these minorities hold an identity relationship to the homeland, or what has been called the land-nurture-identity link (Schwartz 2006). For instance, biodiversity and organic agricultural products are a concern of the German minority in Denmark. With diversification, the German-speaking farmers took the lead in bringing Danish agriculture into the organic sphere and the German association of farmers in Denmark, the *Landwirtschaflicher Hauptverein für Nordschelswig* (LHN), is now the leading organic farming association and represents some of the most important organic farms in Denmark.[5] Moreover, the German minority is also taking a lead in bioenergy production. A bioenergy plant in southern Denmark was initiated by, and is headed by, a member of the German minority.

Similarly, just south of the Danish border, in Schleswig-Holstein, the local Danish minority is taking another tactic with regard to biodiversity. Frustrated by the local government's lack of attention to the environment, a handful of members of the Danish minority have created a grass-roots organization called *Glokale Sydslesvigere* (glocals from South Schleswig) (Dürkop 2007). This action group has been spurred on by the slogan of many environmental activists, 'think global, act local' and their slogan is '*handle/forvandle*', meaning act and change.[6] The newly founded environmental wing even advocates Green budgets and the Green administration of the minority's own affairs. In the same part of the world, a Frisian minority who are among the Frisians living mainly along the coast of north-west Europe has assumed a leading role in protecting the biodiversity of the Wadden Sea around the islands off the coast and in cleaning beaches. These are small national minorities, but they represent movements that populate a space which is usually the domain only of conscientious majorities. The drive to environmentalism of these minorities is motivated not by a desire to hold political power over the territory of their homeland, but by a perceived need to protect and replenish the soil that sustains their collective identity. In other words, it is a politics of regional identity approach to environmentalism.

The Baltic cases

It could also be argued that regional identity informed the Green activism of the titular peoples of the pre-independent Baltic republics. While all three peoples had national-territorial autonomy within the Soviet constitutional framework, they had very little say in managing their own affairs. As is well known, dissent and activism was not sanctioned by the Soviet regime until Gorbachev instigated the policies of *glasnost'* and *perestroika* in the late 1980s. Green activism had understandably been spurred on by the Chernobyl accident in 1985, which also led to the international community paying more attention to environmental degradation in the Soviet Union. In the Baltic republics, those members of the political community drawn from the 'titular' nationalities quickly understood that 'Green' dissent would not meet with the customary Soviet persecution. There were ample opportunities to point out to the regime and the world that the environment in the Baltic republics was degraded. In Estonia it was oil-shale pollution, nuclear risk and mineral (phosphate) mining that motivated the titular people to adopt the environmental discourse (Hosking 1992, p. 181). In Latvia it was the potential damages to the natural environment by large hydro-dams (Daugava River) as well as concerns for tree species as natural symbols of Latvia and the disappearance in general of the agricultural sector (Galbreath & Auers 2009; Hosking 1992, p. 10). As early as the mid-1970s the so-called Great Tree Liberation Movement to register great trees as protected national monuments had emerged (Schwartz 2006). In Lithuania it was nuclear risk as a result of the world's largest nuclear plant being in operation there (Hosking 1992, p. 180). During this period Green political parties emerged in Latvia in 1990 and in Lithuania in 1989, while Estonia had its first Green movement in 1986. While protecting and replenishing the soil that sustained their collective identity was initially the aim of the titular peoples of the Baltic states, the historical circumstances dictated a different road. What might have started as a regional identity approach to environmentalism became an environmental justice approach to nation-building. The difference in the tactics of the titular peoples of the Baltic states and national minorities elsewhere in Europe is therefore significant.

Two discourses, one aim

Whereas national minorities do not enlist history to argue for a return to a previous political status, the titular peoples in the Baltic had historical reasons to believe that a return to the independence experienced between 1920 and 1940 would allow them to restore and rebuild the homeland which nurtured their identity. To win such freedom, they would need to draw on a number of discourses. And, because of *glasnost'* and *perestroika*, such discourses were already in action. Simultaneously with the environmental movements, organizations protesting human rights abuses were also active. Moreover, some of the first instances of dissent were in fact focused on the right to minority languages and culture, rights which had at the time become enshrined in Article 27 of the United Nations' International Covenant on Civil and Political Rights (ICCPR) adopted in 1966. The step from protecting titular languages and culture as a minority right to protecting these as a nation-building cause was not difficult to take. For instance, with the change to multi-party systems, it became possible in both Estonia and Latvia to ensure the adoption of the titular (minority) languages as state languages during 1988–1989 (Aves 1992, p. 35). Movements seeking to protect the titular (minority) nations quickly emerged, often using the environmental card. This was the case in the example of the Estonian Heritage Society, which had been established in 1987. While it had the aim of protecting national (minority/titular) heritage in terms of culture, many of the sites were perceived as supporting the political notion of national (minority/titular) identity. At the same time a group of Estonian economists had put together a proposal, the so-called IME (short for *Isemajandav Eesti* – Economically Self-Governing Estonia), which argued for the well-being of the environment through self-governing (Hosking 1992, p. 183). In Latvia, where ethnic tensions were more pronounced, the justice discourse was less covert than in the other Baltic states. The founding program of the Latvian Popular Front (founded in 1988) focused more on the rights of the titular people than on the environment (see Aves 1992, pp. 31–2). In Lithuania, a more homogeneous republic with less ethnic tension, the environmental and the justice discourses were already occupying the same public space at this time. The leading members of the Lithuanian Popular Front (also established in 1988) consisted of both radical rights and independence activists as well as members of the Green movement (see Aves 1992, p. 33). Inevitably such undertakings became politicized, and by 1988 movements in all three republics had begun to display the national flag at their events.

The environmental and the justice/human rights discourses became a 'protect-our-nation' discourse.

The lines between *protecting* and *promoting* the nation can easily become blurred, as is evident in the case of the Estonian Heritage Society. As nation protection becomes nation-building, the need to seek sovereignty is created and drawing on the nationalistic discourse becomes helpful. Hence, the cause no longer appears environmental, because sovereignty jeopardizes the global aim of environmentalism. Nation-building goes against the cosmopolitan ethos, especially the need to cooperate across state boundaries. The so-called 'beyond sovereignty' (Hurrell 2006, p. 167) environmental approach would violate the political need for state sovereignty and national self-determination. At such a point, liberal and environmental justice usually part ways. However, the ideals of liberal justice were well known to the intellectual dissenters in the Baltic states. Since the Soviet regime was perceived as unjust, the model of justice on offer in the West was more appealing. While it is beyond the scope of this essay to define liberal justice in its many variations, it should be taken into account that the notion of liberal justice is about the well-being not only of individuals but also of human communities. During the second half of the twentieth century, liberal justice had become synonymous with acknowledging the rights of cultural minority communities within national states. This had been established not only in international human rights law, as noted in Article 27 of the ICCPR, but also in the rapidly expanding theory of the modern national state as the home of many diverse cultures, many of which enjoy the right to self-government, as is the case in Western Europe (South Tyrol, Catalonia, the Basque Country).[7] There is thus a need to understand liberal justice in a collective sense. Moreover, there is a difference between collective liberal justice and collective environmental justice. The latter protects all collective entities, including non-human species, whereas the former aims at protecting only human species. This is why the liberal justice discourse can be enlisted not only for the individual but also for the collective by people who seek self-determination and sovereignty. Environmental justice can also be enlisted to seek self-determination and self-administration over soil and territory but would not be acceptable to Green theory if applied to achieve sovereignty, because sovereignty jeopardizes the global aim of environmentalism. Finally, liberal justice often involves the nation, whereas environmental justice is more likely to be regional. This is because territory in environmental justice means biosphere, whereas territory in liberal justice means the sovereign (political) nation. On the collective liberal notion, the nation must therefore be protected against the injustice it is facing.

Injustice can mean violence against the people of the nation as well as the territory of the nation. Territory, including the environmental state of the territory, plays a significant role in both liberal and environmental justice. Territory both sustains people's identity and excludes other identities (Connolly 1995, p. xxii). To occupy territory is to receive sustenance, and to be territorialized is to be occupied by a particular territorial identity. Thus, as the environmental health of the territory is a vital part of people's self-identification processes, both liberal and environmental justice may be called upon to protect it. The preservation as well as the survival of the territory is therefore necessary in order to secure the well-being of both people and nation. Unlike Green justice theory, the titular peoples of the Baltic states were able to engage both the environmental and the liberal justice discourses for the same purpose, the purpose of nation-building.

Towards nation-building

It could be questioned, however, whether at this point the Baltic environmental movements were in fact motivated by minority identities, even though they employed minority strategies. Hosking speaks of the 'ethnic depression' which inflicted many among the titular peoples of the Baltic republics during the Soviet era.

> As tiny nations in a huge empire, without external support and facing a regime which permitted no organised opposition, the Estonians, Latvians and Lithuanians seemed to accept their lot passively, at least after the end of guerilla resistance in the early 1950s. There were, it is true, occasional demonstrations of protest . . . and the odd open letter was smuggled out, but few outside observers thought they represented more than the reaction of a limited circle of intellectuals. As later became clear (and should have been apparent at the time), these manifestations of dissent actually gave vent to the suppressed feelings of whole peoples who could see their languages, their cultures, their customs and their natural environment gradually being undermined or degraded until the question arose whether they could go on surviving as nations at all. (1992, p. 181)

It would appear that this latent depression was in fact national as well as ethnic: this is to say that it drew upon an aspiration for some sort of self-government as opposed to what most ethnic groups desire – the right to preserve and practice their own cultural, linguistic and religious identities without interfering with the business of the state. The members of the new movements were scientists, writers and other intellectuals who feared for the future of their nations and national cultures under the Soviet regime (Hosking 1992, pp. 181–2). The huge immigration to Estonia and Latvia during the Soviet period had in both cases decreased the percentages of the local population drawn from the titular national group. Intellectuals felt that the Soviet immigration and the Russian language were encroaching on the survival of their nations. The fact that the titular people might be seen as an ethnic group rather than a nation within the Soviet Empire was not accepted easily by intellectual leaders in these republics. The leaders of both the environmental and the justice discourses, whose identities had been territorialized during the Soviet regime and in opposition to the Soviet regime, reacted therefore by becoming more national and less ethnic.

The ethnification-cum-nationalizing of both discourses, especially in Estonia and Latvia and less so in Lithuania, is an important aspect of the evaluation of the environmental discourse that took place in the Baltic republics prior to independence. While the degree of ethnic divide within these societies determined the extent to which the environmental discourse was overt and the justice discourse initially covert, as noted earlier, it seems clear that ethno-national mobilization was happening even as the environment was still on the agenda. Certainly, the environmental discourse was becoming an instrumental tool of liberal justice. But rather than wanting justice for the sake of the environment, the titular peoples of the Baltic republics wanted justice for the sake of the health of the nation. Contrary to other post-communist states, the titular peoples of the Baltic states had strong memories of their own nation.[8] Moreover, having briefly experienced the freedoms of Western liberal justice in the years between the two wars, the elites of the new movements were probably well versed in the liberal notions of the right of self-determination of peoples as defined by modern international law and popular sovereignty as developed in liberal political thought.[9] In short, ethnic meant national and justice therefore came to mean national self-determination and state sovereignty.

As independence was achieved, nation-building started in earnest. It is no truism that nation-building and nationalism have gone hand in hand for centuries in most of the liberal West. Thus, as has been reported widely, the final steps towards independence which had involved a rise in ethno-nationalist rhetoric gave way to full blown nationalism after independence (Galbreath & Auers 2009; Hosking *et al.* 1992). Environmentalism thus waned even at a time when in Latvia the Green Party was able to participate in coalition governments and at one point held the prime minister's office. What had happened was a fusion or 'discursive overlap' (Galbreath & Auers 2009) of environmental justice politics and nation-building politics – or interest politics and ideological politics.

The move to ideological politics of nationalism in the post-independence Baltic states was arguably inevitable given the historical circumstances. Whether one can speak of nationalism 'mining' the environmental discourse, as has been suggested (Galbreath & Auers 2009), or whether nationalism in fact fused with environmentalism is the question. The mining theory – meaning the moral borrowing of environmental arguments to legitimize the nationalist agenda – makes sense if we approach the question from the perspective of rational interest politics. In that case the people who did the mining were consciously trying to 'draft' the environmental approach into the nation-building one. This would mean a strategically and rationally framed approach. In contrast, from the perspective of discourse analysis and identity politics, it may be feasible to argue that actors within the environmental justice discourse, i.e. the leaders of the new movements, co-opted the liberal ideology of self-determination and popular sovereignty as indicated above. In their rhetoric, the lines between environmentalism and nationalism in terms of nation-building became blurred. They no longer saw environmental justice as an adequate politics to protect the territory that gave them sustenance and identity. Instead they saw liberal justice and ideology as the best means to protect their territory and thus their nation. The overlap of nation-building and state-building in liberal ideology since the French Revolution means that national identity is derived from a certain national territory. The environmental protection of such national territory is therefore an intrinsic part of both national identity and the nationalism that fosters such identity. In the final section, I try to analyze this problem in order to determine whether the nationalistic approach that was adopted did in fact retain some Green elements.

Eco-nationalism and Green citizenship

As already noted, de-Shalit finds three positive aspects about nationalism that speak to the ideology of ecologism. It may foster solidarity and responsibility towards fellow citizens, it may promote preservation of national heritages, and it may foster a sense of obligation to future generations (de-Shalit 2006). If these are acceptable ideals to Green theory, the argument that environmentalism and nationalism are not compatible seems erroneous. Taken one by one, there might be a reason to argue for Baltic eco-nationalism. First, it could be argued that the concern in Estonia with oil-shale pollution was clearly evidence of responsibility towards fellow citizens. The concern with nuclear risk may also be seen as responsibility towards fellow citizens although it could be argued that ultimately it is probably more driven by self-interest. On the other hand, the concern with mineral (phosphate) mining was an example of lack of solidarity, since the movement was to a great extent concerned with the arrival in Estonia of more Russian labor which might tip the majority ratio towards the Russian part of the population. The concern in Latvia about potential damage to the natural environment by large hydro-dams (Daugava River) appears as a sign of responsibility not only towards the environment but also towards fellow citizens who would be affected by this. In Lithuania nuclear risk was the primary cause of concern given that the world's largest plant was on Lithuanian territory. As noted, this could result in both responsibility towards fellow citizens and self-interest. Second, the notion that nationalism may promote the preservation of national heritages applies in both Estonia and Latvia. In Estonia the aforementioned Heritage Society worked especially in rural areas to protect national cultural sites. In Latvia the Great Tree Liberation Movement sought to protect great trees as national monuments. The desire in all three republics to protect the titular languages through the adoption of legislation to make them official languages is clearly an example of preservation of national heritage unless one holds an instrumental view of language. Third, the idea that nationalism may foster a sense of obligations to future generations is evidenced not only in the nuclear-risk concerns but also by the concern in Latvia with the disappearance of the agricultural sector; a problem which many national states in Europe had to deal with after 1945. Clearly, the risk of nuclear disaster is arguably a concern if future generations are to exist and live healthy lives. But the survival of an agricultural sector is also a concern not only to the national state but also to the global survival of species. Without agriculture future generations may not eat as healthily as today. While neither exhaustive nor perfectly analyzed, these are factors that seem to speak to eco-nationalism.

Furthermore, all three Baltic states elected soon after independence to apply for membership of the European Union (EU). This would indicate a willingness not only to relinquish some sovereignty, but also to implement the Green policies that are dictated from Brussels. Moreover, the governments of the three states have continued to implement environmental projects, of course with more or less success (Schwartz 2006). And as noted, Latvia had a Green coalition government for a while (Galbreath & Auers 2009). Thus, the Baltic scenario would seem to indicate that there are some mitigating forces that allow nation-builders to be environmentalists while also seeking to protect the sovereignty of their nations. Of course, the desire in the Baltic states to join the 'club of Europe' as soon as possible in order to distance themselves further from the former Soviet Union should not be neglected. Whether rational or emotional, the centrifugal and the centripetal forces were at work at the same time.

This may indicate that environmentalism and nationalism could cohabit in the politics of the Baltic states after independence.

This is not to argue against the view that at times eco-nationalism appears rather more 'Brown' than Green, meaning ethnocentric (Galbreath & Auers 2009). Certainly, nation-building in the three Baltic states got off to a rocky start in terms of normative frameworks skewed in favor of the titular peoples, and Estonia and Latvia still have a way to go in terms of political and social integration of the Russian-speaking populations (Malloy, forthcoming). The language laws adopted after independence are a case in point. Nevertheless, it could be argued that Green goals are ethical goals no matter what motivation because they are universal, meaning they involve all humankind. Green universal goals are thus morally correct because they seek to preserve the globe for all as they do not recognize boundaries, whether sovereign, ethnic or linguistic. In my mind it therefore comes down to what type of citizenship was informing the environmental activists in the Baltic states at the time.

Dobson argues for two types of Green citizenship (2003). The difference between the two lies to a great extent in the ability of each to accommodate the rights discourse. While the environmental notion of citizenship is often an extension of the rights discourse, the right not to be subjected to environmental risks and threats, the ecological notion emphasizes responsibility and virtues.[10] Dobson thus defines ecological citizenship as a concept that

> . . . deals in the currency of non-contractual responsibility, it inhabits the private as well as the public sphere, it refers to the source rather than the nature of responsibility to determine what count as citizenship virtues, it works with the language of virtues, and it is explicitly non-territorial. (Dobson 2003, p. 89)

The two notions of citizenship are not counterproductive, according to Dobson; rather they are complementary in that they both aim to make our planet environmentally sustainable. In terms of citizenship theory, ecological citizenship is in a separate category because it rejects a number of liberal ideals, most specifically the liberal notion of neutrality towards the private sphere of people's lives. In order to create law-abiding, responsible Green citizens, the ecological version of citizenship must reach into the private sphere where much environmental decay is generated. Environmental citizenship is, on the other hand, an expansion of the traditional three-fold view of liberal citizenship that holds that as citizens we have certain civil, political and social rights (Marshall 1950). A right not to be subjected to environmental risks and threats would constitute the fourth dimension of that concept (Steenbergen 1994, p. 142, in Dobson 2003, p. 87).

On this dichotomist view of Green citizenship, the environmental actions of the titular peoples in the Baltic republics before independence could be seen as a sign of an emerging environmental citizenship. The right to environmental justice in terms of environmental rights was clearly at the forefront of this, as suggested in the discussion above. The Soviet regime paid little or no attention to environmental rights, and only began to address environmental degradation when the international community reacted. Thus, the desire to manage the health of their territory better drove the titular peoples towards environmental justice and perhaps environmental citizenship. But at the same time as the titular peoples were Green citizens they were also eco-nationals; they were proponents of the three positive notions of nationalism that speak to Green theory. These notions of eco-nationalism are considered morally correct because they involve all members of society. The meeting of environmental justice and eco-national ethics weighs in favor of an argument for eco-nationalism. Thus, the co-optation of liberal justice by the environmental discourse may in fact have formed the bridge to eco-nationalism.

Conclusions

The modest portrait of Green theory that I have offered in this essay could well be summed up as a new social movement seeking to define a Green citizenship. There are many reasons why the minority environmentalism discussed in this essay may be seen as environmental citizenship in the making. Unlike many members of majorities, territorial minorities see the necessity of protecting the Earth precisely because they derive great sustenance from the territory that they inhabit. The minorities who enlisted the environmental discourse and sought self-determination over the environmental management of their homelands evidenced both Green ethics and Green ideology in this politics. First and foremost, they promoted the core tenet of bioregionalism that Green theorists consider a strong element in the quest for the behavioral change of human beings. The community is where this is best promoted, and minorities often live in well-defined communities. Moreover, Green sustainable agriculture is considered a cornerstone of ecologism which is also best pursued in communities of the bioregions. Many territorial minorities support themselves primarily through farming. Similarly, Green theory argues that behavioral change may best be promoted within certain groups, such as certain social classes. Here territorial minorities are also in the position to deliver, as they usually constitute a cohesive group, if not by class but by status, race or ethnicity. Thus the revolution in practices that Green theory calls for is actually easily supported by minorities because to them it means freedom – freedom to manage their own affairs and protect their homelands. In protecting their homelands, the minorities furthermore take responsibility for the stewardship of the territory, and they show solidarity towards their fellow citizens. In seeking the ultimate goal of ecologism, the survival and flourishing of the region, minorities also show obligations to future generations. In short, territorial minorities practicing environmental citizenship contribute to the transformation that Green theorists advocate.

The complication to this argument as far as the environmentalism of the titular peoples of the Baltic republics and the nationalism of post-independence are concerned is that environmentalism does not accommodate sovereignty well. When history handed the titular peoples sovereignty rather suddenly and without bloodshed, it was not, however, without cost. The territory of sovereignty that they inherited was not 'pure' in terms of the territorial identity it sustained. The large percentages of ethnic Russians living in these new sovereign states made it difficult for the titular peoples to create a pure (Green) national identity; hence, the intensification of the nationalistic discourse which had already taken form during the last years of Soviet regime. While at times in the post-independence Baltic states the nationalistic discourse may be stronger than the environmental one, I believe I have demonstrated that Greens and nationalists came together both in theory and in practice prior to independence. My argument is therefore that notwithstanding post-independence xenophobic and nationalistic rhetoric, including that of Green politicians, Green goals cannot be discarded the way other political goals may be. Green goals do not easily disappear because time will remind nationalists that protection of the environment only becomes more urgent as the years move on. At best Green goals can be neglected for a while. Hence, as the post-independence discourse develops further, Green goals are likely to resurface from time to time. It may simply be premature to judge the titular peoples only on the nationalistic discourse. As the nations in the Baltic states consolidate, the titular peoples may well have an environmental foundation for Green citizenship. The pre-independence activism discussed in this essay would seem to indicate that the Green nationalists of today might become the national Greens of tomorrow.

Notes

1 In the following I will use the term minority to refer to national and ethnic minorities as well as indigenous peoples, except where specifically naming distinct groups is necessary for clearer understanding.

2 I recognize that the titular peoples of the Baltic republics were not minorities according to the Soviet constitution. I use the term here, however, since their approach to action in environmental politics was similar to that of minorities in politics in general.

3 The body of green political thought is very diverse, and it would be impossible within the scope of this essay to pay attention to all theories offered. I will therefore, with a few exceptions, refer to green political theorists with little discrimination as to which green theorists promote what theories. This unfortunate simplification is clearly only with the one purpose of making it accessible to the non-green reader. For any unwarranted mistreatment of presentations and arguments, I sincerely apologize.

4 Derek Bell argues for 'green liberalist liberalism' based on a compatibility of Rawls's political liberalism and environmentalism (2002). Mary Mellor similarly argues a strong case for green socialism (2006).

5 See website, http://www.lhn.dk/, accessed 28 May 2009.

6 See website, http://www.forvandle.de/.

7 Will Kymlicka (1995) presents the best overview and the best examples of this development.

8 Shari Cohen has argued that a profound ideological vacuum fueled destructive tension throughout post-communist Europe (1999). Using Slovakia as an example of this historical amnesia she argues that communist socialization prevented people from tying their private stories to a larger historical narrative. Her analysis is probably correct with regard to Slovakia, but in my mind could not be applied to post-independence Baltic states.

9 While popular sovereignty became part of liberal ideology with the French Revolution, the right of self-determination of peoples is enshrined in Article 1 of the ICCPR.

10 For a critique of Dobson's arguments see Hayward (2006).

References

Aves, J. (1992) 'The Evolution of Independent Political Movements after 1988', in Hosking, G. A., Aves, J. & Duncan, P. J. S. (eds) (1992), pp. 29–66.

Bahro, R. (1986) *Building the Green Movement* (London, GMP).

Ball, T. (2006) 'Democracy', in Dobson, A. & Eckersley, R. (eds) (2006).

Barry, J. (1993) 'The State (and Citizen) of Green Political Theory', Paper given at Annual Conference of the Political Studies Association, University of Swansea, 29–31 March.

Becker, M. (2006) 'Rhizomatic Resistance: the Zapatistas and the Earth Liberation Front', *Green Theory & Praxis: The Journal of Ecopedagogy*, 2, 2, available at: http://greentheoryandpraxis.csufresno.edu/articles/becker.doc, accessed 27 January 2009.

Bell, D. (2002) 'How can Political Liberals be Environmentalists?', *Political Studies*, 50, 4, pp. 703–724.

Cohen, S. J. (1999) *Politics Without a Past: The Absence of History in Postcommunist Nationalism* (Durham, NC, Duke University Press).

Dagger, R. (2006) 'Freedom and Rights', in Dobson, A. & Eckersley, R. (eds) (2006).

Dawson, J. I. (1996) *Eco-nationalism: Anti-nuclear Activism and National Identity in Russia* (Durham, NC, Duke University Press).

de-Shalit, A. (2006) 'Nationalism', in Dobson, A. & Eckersley, R. (eds) (2006).

de-Shalit, A. (2006) 'Thirty Years of Environmental Theory: From Value Theory and Meta-Ethics to Political Theory', *Critical Review of International Social and Political Philosophy*, 9, 1, pp. 85–105.

Dobson, A. (1990) *Green Political Thought,* 3rd ed. (London, Routledge).

Dobson, A. (1996) 'Representative Democracy and the Environment', in Lafferty, W. M. & Meadowcroft, J. (eds) (1996), p. 137.

Dobson, A. (1998) *Justice and the Environment. Conceptions of Environmental Sustainability and Dimensions of Social Justice* (Oxford, Oxford University Press).

Dobson, A. (2003) *Citizenship and the Environment* (Oxford, Oxford University Press).

Dobson, A. (2006) 'Citizenship', in Dobson, A. & Eckersley, R. (eds) (2006).

Dobson, A. & Eckersley, R. (eds) (2006), *Political Theory and the Ecological Change* (Cambridge, Cambridge University Press).

Dryzek, J. (2000) *Deliberative Democracy and Beyond* (Oxford, Oxford University Press).

Dryzek, J., Honig, B. & Phillips, A. (eds) (2006) *The Oxford Handbook of Political Theory* (Oxford, Oxford University Press).

Dürkop, D. (2007) 'Sydslesvig skal spare energy. Miljø. Græsrødder vil sætte miljøet på den sydslevigske dagsorden', *Flensborg Avis*, 31 March, p. 7, available at: http://www.flensborg-avis.de/index.php?c=v2124394.339, accessed 28 May 2009.

Eckersley, R. (1987) 'Green politics: a practice in search of a theory', Paper delivered at Ecopolitics II Conference, University of Tasmania, 22–25 May.

Eckersley, R. (2006) 'Communitarianism', in Dobson, A. & Eckersley, R. (eds) (2006).

Elliot, R. & Gare, A. (eds) (1983) *Environmental Philosophy* (Milton Keynes, Open University Press).

Galbreath, D. J. (2005) *Nation-Building and Minority Politics in Post-Socialist States. Interests, Influences and Identities in Estonia and Latvia* (Stuttgart, ibidem-Verlag).

Galbreath, D. J. & Auers, D. (2009) 'Green, Black and Brown: Uncovering Latvia's Environmental Politics', *Journal of Baltic Studies*, 40, 3.

Goldsmith, E. (1972) *A Blueprint for Survival* (London, Tom Stacey).

Goodin, R. (1992) *Green Political Theory* (Cambridge, Cambridge University Press).
Hayward, T. (1994) 'The Meaning of Political Ecology', *Radical Philosophy*, 66, Spring, p. 163.
Hayward, T. (1998) *Political Theory and Ecological Values* (Cambridge, Polity Press).
Hayward, T. (2006) 'Ecological Citizenship: Justice, Rights and the Virtue of Resourcefulness', *Environmental Politics*, 15, 3, pp. 435–446.
Hosking, G. A., Aves, J. & Duncan, P. J. S. (1992) *The Road to Post-Communism. Independent Political Movements in the Soviet Union 1985–1991* (London, Pinter).
Kenny, M. (2004) *The Politics of Identity* (Cambridge, Polity).
Kumar, K. (ed.) (1984) *The Schumacher Lectures*: Vol. II (London, Blond & Briggs).
Kymlicka, W. (1995) *Multicultural Citizenship* (Oxford, Oxford University Press).
Lafferty, W. M. & Meadowcroft, J. (eds) (1996) *Democracy and the Environment* (Cheltenham, Edward Elgar).
Lønborg, B. (2009) 'Skisport på miljøets betingelser' ['Skiing on green conditions'], *Flensborg Avis*, 3 January, pp. 20–1, available at: http://www.flensborg-avis.de/, accessed 28 May 2009.
Malloy, T. H. (2007) 'Are there any Green minorities?', paper given at the 4th ECPR General Conference, Pisa, 6–8 September.
Malloy, T. H. (forthcoming) 'Social Cohesion Estonian Style: Minority Integration through Constitutionalized Hegemony and Fictive Pluralism', in Agarin, T. & Brosig, M. (eds) (forthcoming, 2009) *Minority Integration: Debating Ethnic Diversity in Eastern Europe* (Amsterdam: Rodopi).
Marshall, T. H. (1950) *Citizenship and Social Class and Other Essays* (Cambridge, Cambridge University Press).
Mellor, M. (2006) 'Socialism', in Dobson, A. & Eckersley, R. (eds) (2006).
Meyer, J. M. (2006) 'Political Theory and the Environment', in Dryzek, J.S., Honig, B. & Phillips, A. (eds) (2006), p. 782.
Naess, A. (1973) 'The Shallow and the Deep, Long-range Ecology Movement. A Summary', *Inquiry*, 16, 1, p. 95.
Sale, K. (1984) 'Mother of All: An Introduction to Bioregionalism', in Kumar, K. (ed.) (1984).
Sale, K. (1985) *Dwellers in the Land: the Bioregional Vision* (San Francisco, Sierra Club).
Salleh, A. (1997) *Ecofeminism as Politics: Nature, Marx and the Postmodern* (London, Zed Books).
Schwartz, K. Z. S. (2006) *Nature and National Identity after Communism: Globalizing the Ethnoscape* (Pittsburgh, University of Pittsburgh).
Smith, M. J. (1998) *Ecologism. Towards Ecological Citizenship* (Buckingham, Open University Press).
Stransky, N. B. (2006) 'The Disaster of Coercive Institutional Socialization: Alaska Natives and Environmental Emergency Planning', *Green Theory & Praxis: The Journal of Ecopedagogy*, 2, 1, available at: http://greentheoryandpraxis.csufresno.edu/articles/AlaskaPaperpublication.doc, accessed 27 January.
Thompson, J. (1983) 'Preservation of Wilderness and the Good Life', in Elliot, R. & Gare, A. (eds) (1983).
Von Steenbergen, B. (ed.) (1994) *The Conditions of Citizenship* (London, Sage).
Wissenburg, M. (2006) 'Liberalism', in Dobson, A. & Eckersley, R. (eds) (2006).

SUSTAINABLE DEVELOPMENT IN LITHUANIA: BETWEEN THE GOVERNMENT AGENDA AND THE UNDISCOVERED CIVIL SOCIETY

Nora Mžavanadze

Introduction

The impact of social capital in the realms of public policy, such as education, public health and the environment, is well documented. However, the role of social capital in the formation and implementation of public policy is a rather unexplored area. A paradigm shift in public policy-making has created more room for social capital in this arena. Modern government is no longer considered to be effective in formulating and implementing public policies on its own; there is a need for new types of societal involvement in policy-making. Therefore, governance, as opposed to government, is a more crowded policy-making arena, working towards achieving common goals instead of simply regulating and controlling. The boundary between the public and private sector has become blurred and the balance has shifted between two extreme paradigms. Everything is left either to the state or to the market.

Public policy officials have been under pressure to utilize business management techniques in public management since the 1980s in order to make the public sector more efficient and effective. The most common definition of strategy describes it as 'top management's plans to attain outcomes consistent with the organization's missions and goals' (Wright *et al.* 1992, p. 3), whereas strategic management is a process of facilitation and implementation of specific measures and actions in order to achieve transition to the desired state. The field of strategic management has undergone a paradigm shift, which carries implications for public policy-making. Mintzberg *et al.* (1998) demonstrate the variety of approaches and the evolution of strategic management. They present ten schools of strategic management; some of them are discussed more as strategy formulation techniques (design, planning, positioning, entrepreneurial and learning schools), others are offered as a theoretical lens for analyzing strategy making (cognitive, power, cultural, environmental and configuration schools). The strategic management paradigm has evolved from being an area of exclusive external expertise to an interactive and engaging learning process incorporating all stakeholders. Consequently, social capital comes into the strategic management field, ranging from zero contribution (in the case of a design school according to Mintzberg *et al.* 1998) to being the key vehicle (in the case of a learning school according to Mintzberg *et al.* 1998).

Sustainable development is a newly emerging public policy area aiming to become a mainstream policy-making arena through a comprehensive learning process across Europe. Although the origin of sustainable development in international policy documents can be traced back to the Rio Summit on Environment and Development in 1992, the concept only entered national public policy documents in 2002. The majority of European countries prepared sustainable development plans, assessments or strategies before or straight after the Johannesburg summit on Sustainable Development in 2002. It is likely that sustainable development policy has just completed or is near to completing a policy cycle of formulation, implementation, monitoring, evaluation and review. A National Strategy for Sustainable Development (NSSD) has become the main framework of sustainable development policy at national level in many EU member states. The concept of sustainable development relies on multi-stakeholder participative consensus-building, which can only be achieved where there is a high level of social capital present. In order to be a mainstream policy, the sustainable development paradigm would need to be the main paradigm of long-term national development plans and would need to be integrated into all other governmental policy sectors, across all levels of jurisdiction. This is not the case anywhere in the world today. Nevertheless, the EU and some other countries are approaching this goal through a continuous process of improvement and learning (Spangenberg 2006).

The EU boasts a range of both informal and binding agreements that address the economy–ecology nexus, most recently the Aarhus Convention. However, national initiatives to address issues at the core of sustainable development appeared in 1992 in the form of Agenda 21 and since then many governments have embarked on this new policy-making paradigm. International organizations (UN, OECD) have become the major driving force of this paradigm worldwide and the European Union (EU) encourages its member states to constantly innovate in this new policy field (Baker 2000). Vague guidelines have been developed for NSSDs to allow adaptation, but these have resulted in a freedom to choose from a variety of approaches towards NSSDs and their outcomes differ as well.

As a consequence, many governments have felt obliged to develop a strategic document. Agenda 21 (1992), however, aimed at improving or restructuring 'the decision-making process so that the consideration of socio-economic and environmental issues is fully integrated and a broader range of public participation assured'

(Agenda 21 1992, paragraph 8.3). Cherp *et al.* (2004) also emphasized that NSSDs should not be seen as new plans, but should start new processes that could improve governance by involving more stakeholders and therefore enhancing social capital.

The aim of this essay is to explain under what conditions social capital facilitates policy-making and its implementation by investigating Lithuania's sustainable development policy. The study starts with a brief summary of the theoretical underpinnings of sustainable development and social capital theory. An analysis of NSSD processes in Lithuania is then carried out through the lens of stakeholder involvement in the process of formulation and implementation. In conclusion, the implications of social capital involvement for policy outcomes are assessed.

Linking social capital and sustainable development

The concepts of social capital and sustainable development appeared simultaneously and are generating an increasing amount of interest in the academic world. They have rarely been considered together, but both carry many common characteristics and are linked at the conceptual level.

Researchers in the fields of social capital and sustainable development have experienced very similar patterns: contested views regarding definition, operationalization, measurement and methods (Adam & Roncevic 2003; Lehtonen 2004). Both concepts aim to become the intersection point for many different disciplines and their application is usually customized for the context. Both concepts have been adopted by international development agencies and have been treated as the 'missing link' and the 'cure for all' in various development and transition challenges (Bebbington *et al.* 2004; Lele 1991); both applications have received criticism (Doyle 1998; Harriss 2002). The only significant difference between the concepts of social capital and sustainable development is that the latter stems from international policies in the field of environment and development and is a practical concept, while the former entered the field of policy-making after being developed on the theoretical level.

Sustainable development was first defined in the Bruntland report in 1987 as 'development that meets the needs of the present without compromising the ability of future generations to meet their own needs' (WCED 1987, p. 43). It is also described as a normative concept resting on three equally important pillars: economic, social and environmental. In simple terms, it means enhancing human development while also respecting environmental limitations. The concept of social capital emphasizes the importance of social networks, shared values, cooperation and trust between humans in the challenge of fitting the needs and wants of humanity into the environmental constraints of our planet.

There is not a defined list of forms of capital for sustainable development, but usually their type is based on three pillars: economic, social and environmental. The social pillar is the most obscure of all since it is the least tangible of the three (Lehtonen 2004). However, according to Woolcock (2001) the social and institutional side of development is gaining more and more attention due to events such as the fall of communism and financial crises as well as the persisting problems of development.

The social pillar of sustainable development usually encompasses a huge variety of issues, starting from poverty alleviation, malnourishment and shelter, followed by inequality and inadequate social services, and culminating in complex issues of institutions, governance, capacities, participation and empowerment.

The concept of social capital enters the sustainability debate as a means to enable institutional and behavioral change for achieving sustainable development. Implementation of sustainable development can be viewed as a collective action problem and social capital can be seen as 'norms and networks that facilitate collective action' (Woolcock 2001, p. 70). Social capital in the context of sustainability is defined as 'networks of social relations characterized by norms of trust and reciprocity that can improve the efficiency of society by facilitating coordinated actions' towards implementation of sustainable development (Lehtonen 2004, p. 204). In this context an assumption is made that social capital enables collective action. However, 'while a large body of research has focused on the benefits of social capital, the literature on its risks is much sparser' (Adler & Kwon 2000, p. 106).

Social capital is linked to all sustainability pillars; economic, social and environmental. It has an impact on the outcomes of the management of natural resources, health prospects, economic development and education, which are also key areas for advancing sustainable development (Agenda 21 1992; Islam *et al.* 2006; Pretty & Ward 2001; Schuller & Field 1998; Woolcock 2001). Therefore, Lehtonen (2004) claims 'a part of the social capital is critical, irreplaceable, in the same manner as some of the natural capital is considered "critical"' (Lehtonen 2004, p. 204).

When analyzing sustainable development initiatives in the light of social capital theory, Devine-Wright *et al.* (2001) start with three key assumptions. First, implementing sustainable development means deviating from 'business as usual'. Second, in order to deviate from 'business as usual' a change in opinion, perception and value of procedures, technologies and organizations must occur which could enable all kinds of innovations. The third assumption speaks about the need for multi-stakeholder dialog, because a single organization or individual cannot perform this task successfully. At the same time, multi-stakeholder participation platforms act as a melting pot for the formation of social capital. Sustainable development embraces many diverse issues and the realization of the complex links between the issues is likely to emerge through the awareness-raising, assessment, deliberation and negotiation processes of NSSDs. Therefore, the role of social capital as networks, trust, reciprocity and institutions is crucial for the implementation of NSSDs.

Understanding the involvement and formation of social capital in the processes of an NSSD emerges as a key aspect of understanding the successes and failures of this public policy. Under what conditions can multi-stakeholder participation be useful for public policy formulation and implementation? Under what conditions can multi-stakeholder processes mobilize the necessary social capital? The role of the government and process characteristics need to be considered carefully before designing the multi-stakeholder participation mechanism for public policy.

A National Strategy for Sustainable Development becomes the main framework for sustainable development policy at the national level. It is defined as a 'co-ordinated set of participatory and continuously improving processes of analysis, debate, capacity-strengthening, planning and investment, which integrates the economic, social and environmental objectives of society, seeking trade-offs where this is not possible' (OECD 2001b, p. 16). These processes eventually materialize into a multi-stakeholder societal consensus document (Steurer & Martinuzzi 2005), which indicates the priority areas, desired goals and the means to achieve them and builds the foundation for an institutionalized long-lasting national initiative towards achieving sustainable development.

The role of the state in the case of NSSDs is multi-faceted. 'The extension of participation does not mean that the role of government is in any case diminished. On the contrary, governance for sustainable development depends first and foremost on active governments that place this goal at the centre of the political agenda' (Meadowcroft 2004, p. 187–8). On one hand the state is the initiator of the NSSDs, facilitator of the multi-stakeholder debate and also the main 'responsibility-bearer' along with its main social partners. On the other hand, the state has levers to influence the quality and quantity of social capital in the country and the outcomes can range from dependent citizens to entrepreneurial citizens (Ostrom 2000). From this point of view, several assumptions underlie the reference to social networking, in particular for building upon social capital for the initiative of an NSSD. Firstly, being a relatively new, complex and 'different' branch of public policy, sustainable development requires the capacity and expertise of public administration (Lafferty 2004). When these are not developed to sufficient levels to deal with all scales and sectors of sustainable development, the engagement of members of society, such as the mobilization of social capital, may be of great help to policy makers. Second, not all societies affected by the policies of sustainable development possess social capital in the form required for effective cooperation with policy-makers (Mihaylova 2004; Rydin & Holman 2004).

The design of public participation processes is a delicate task, which should be subject to public scrutiny. There are many ways to involve the public in policy-making ranging from simply informing the public to empowering them in decision-making activities. On the whole, it is important to find a balance between efficiency and effectiveness of public participation. This challenging project may easily turn into a time consuming, costly and ineffective activity; 'however, the choice is usually not whether there should be public participation, but rather about what form participation should take' (Meadowcroft 2004, p. 186). The benefits of public participation include better decisions, better implementation, enhanced legitimacy, transparency, accountability, trust and fairness for the government, better opportunities for learning, self-fulfillment and services of public policy for the wider public (OECD 2001a). OECD (2001a) provides general guideline-principles for successful information, consultation and exchange between the stakeholders in policy-making. This can range from the engagement of public administration to citizen participation, as well as adequacy in resource allocation, government support for capacity building and free information flow.

The 'differentness' of sustainable development also dictates specific requirements for public participation. Meadowcroft (2004, pp. 166–7) argues that public participation processes for sustainable development should allow adequate representation of implicated interests and openness to public scrutiny, deliberative engagement among the implicated parties, the application and integration of different forms of knowledge to decision making and the promotion of societal learning. Were these means embraced by the Lithuanian policy makers during the formulation of the Lithuanian NSSD? I will investigate this in the next section of the essay.

Finally, NSSDs have another important role in the former socialist states of Central and Eastern Europe. If applied properly; they could be tools for the enhancement of democracy. According to *The Economist* democracy index (Kekic 2007), most former socialist countries of Central and Eastern Europe are ranked as flawed democracies due to a lack of political participation and political culture. Although the 'old' and the 'new' Europe have formally declared almost the same level of political freedoms and civil liberties, their realization still lags behind in 'new Europe'. A greater need for transparency, accountability and trust is also felt. Former socialist countries are still exposed to a narrow, expertise-based policy-making paradigm coming from the recent socialist past. The democratic tradition of open, transparent and participative policy-making has not yet become a tradition and a major value change has to happen in society to make this a reality. Therefore, the framework of NSSDs in the form of international guidelines is probably one of the first exposures to a participatory, democratic and engaging policy-making system for former socialist countries.

The role of public participation in policy-making: the case of Lithuania's NSSD

The formulation of the Lithuanian NSSD started in 2000 with the establishment of Commission on Sustainable Development, consisting primarily of state officials. In 2002 a comprehensive report on the implementation of sustainable development in Lithuania was issued specifically for the Johannesburg Summit. A total of 37 state officials and academics prepared the document (MoEoL 2002).

Following this, a National Strategy for Sustainable Development was formulated on the basis of the Johannesburg report in 2003 (NSSDoL 2003). A group of nine state officials and academics produced the strategy. Preparation of both documents was financially supported by the United Nations Development Programme office in Lithuania (UNDP 2005). The Ministry of the Environment was assigned the task of coordination and development of the NSSD. In 2005, another report on implementation of the NSSD was prepared by 11 state officials and academics (MoEoL 2005). In 2006, the first review process was initiated. The review panel was formed mainly of state officials and academics.

The Lithuanian NSSD is an example of a case with non-existent public participation, which is mainly limited to ministerial and scientific community collaboration. How could it happen that one of the essential elements of the NSSD process was omitted despite international and EU guidelines? Initially the public officials and academics responsible for formulating the strategy tried to engage the public in the process. However, as in most of the European experiences with NSSDs, they were disappointed by the modest inputs of public participation. A possible explanation could be the small amount of time given to the production of the document, although, as the Slovak experience shows, even a two-year-long process of strategy formulation with extensive public participation is not enough (Steurer & Martinuzzi 2005).

Nevertheless, the reasons for non-existent public participation procedures are much deeper and more sophisticated than just technical timing difficulties. After dozens of years of socialist non-participatory bureaucratic planning, Lithuanian public administration was given the task of launching a new inter-sectoral and multi-level policy. No dialog between different stakeholders had previously taken place in Lithuania (Mačiulytė & Ragauskas 2007). Over the last two decades participatory policies implemented in Western European states became a norm as a result of the continuous exchange between policy-makers and the public, but in Lithuania the concept of multi-stakeholder and participative policy-making is still not fully embraced. Neither the public administration nor their diverse stakeholders were willing to cooperate in the case of the Lithuanian NSSD. Policy-making in Lithuania is often still considered a narrow expertise issue, where the public administration may collaborate only with academia. From the government's perspective, significant barriers exist for the civil society's ability to influence the government even at a municipal or local level (Žiliūkaitė *et al.* 2006). A comparative World Bank study on strategic planning and policy-making in Latvia and Lithuania has also discovered that Lithuanian policy-making is rather closed, rigid and statutory, whereas Latvia 'allows for an unusually welcome degree of involvement for "outsiders" in the policy process' (World Bank 2006, p. 52).

As a result of the rigid expertise-based policy-making, a slowly emerging civil society is not given enough opportunity to contribute to policy formulation and implementation. This further hinders the development of professional and active civil society. From another perspective, the Soviet heritage still persists in the general social perceptions towards activism. It is still considered to be suspicious, untrustworthy and decadent even today. This passivity and negative attitude towards activism is still deeply rooted in the mentality and hinders the enhancement of democracy. Žiliūkaitė *et al.* (2006, p. 275) diagnose that 'Lithuanian society is suffering from the syndrome of impotence . . . the most severe civil disability which hinders the development of civic initiatives in Lithuania is the society's prevailing disbelief that citizens' collective action can make a difference or help achieve significant outcomes'. Interestingly, this disbelief according to Žiliūkaitė *et al.* (2006) is not based on the actual experience of trying to influence decision-making. 'Collective action through formal organizations and informal networks in Lithuania is still extremely scarce' (Žiliūkaitė *et al.* 2006, p. 276).

Public participation is emphasized in all EU guidelines and communication to the member states concerning NSSD formulation and further processes (EUCOM 2005a, 2005b). However, it seems that the complex task of the NSSD in Lithuania has been entrusted to the country's professionals. The public and multi-stakeholder participation element has also been removed in the monitoring, reporting and review stages of NSSD.

The contents and implementation of Lithuania's NSSD

Non-existent public participation is the main flaw of the Lithuanian NSSD, and also permeates various other aspects affecting the effectiveness of policy. In this section the contents of the Lithuanian NSSD are analyzed and then the impact on policy implementation is assessed.

The literature suggests a variety of approaches to analyzing successes and failures of NSSD implementation. One of the ways to evaluate public policy is to compare the policy outcomes to the initial requirements for the process and contents, which in the case of NSSDs are provided primarily by the UN (UNDESA 2002) and OECD (2001b). The European Union provides guidance for its member states on key principles for NSSD processes and contents (EUCOM 2005a, 2005b) and initiates review procedures, supplying a framework for peer-review analysis (IIEP 2006). There is also a framework for analysis of NSSDs in place, provided by Cherp *et al.* (2004), while Steurer and Martinuzzi (2005) simply examine the five crucial components of the national sustainable development policy: public participation, horizontal policy integration, vertical policy integration, policy implementation and monitoring. The latter approach will be employed to cover the four remaining policy components as public participation has already been covered.

Impact on the contents: a limited vision

The main objective of the Lithuanian NSSD was formulated as follows: 'to achieve the present economic, social and eco-efficiency level of EU-15 countries until 2020, and to secure EU environmental standards' (Juknys 2006, p. 210). This statement assumes that the situation in the more developed European West is better in terms of sustainable development. It also ignores the advantages that the Eastern EU members have inherited from the past, such as richer biodiversity, higher forest density, better quality of forest ecosystems, better and wider use of public transport infrastructure, lower municipal waste accumulation rates and lower energy consumption per capita (VanDeveer & Carmin 2005). The situation is rapidly changing and some indicators are coming into line with Western Europe, but the outcomes may still be influenced to a certain extent. However, the current situation of the EU-15 at the time of writing is not sustainable, hence the EU still requires a sustainable development strategy. Therefore, the current goal of the Lithuanian NSSD appears impracticable.

Furthermore, the strategy continues to present the principles of sustainable development and the main priorities for Lithuania. The list of principles neglects the gender-equality principle, where there is room for improvement. One of the principles calls for a strong state role: 'It is impossible to ensure the right tendency of sustainable development processes . . . without fairly strong management at the state, regional and municipal levels' (NSSDoL 2003, p. 8). This illustrates that Lithuanian policy-making has not made a transition from the strong government paradigm to the governance paradigm.

A Strengths, Weaknesses, Opportunities and Threats (SWOT) analysis was later carried out for sectoral areas representing the three pillars of sustainable development. There is still too much emphasis on 'sectoralism' and too little cross-sectoral consideration. The analysis should concentrate more on trade-offs and synergies between the sectors. Some of the issues considered in the SWOT analysis are outdated due to the rapidly changing economic and social situation (e.g. unemployment). Some issues are being considered in a conventional and technocratic way. For instance, the transport section considers conventional means of problem-solving, such building and widening more roads, bridges, bypasses and viaducts (NSSDoL 2003). It thus ignores the more sustainable means of getting about, such as public transport, cycling and walking.

As a narrow public administration expertise issue, the NSSD fails to encompass emerging national development issues. It embraces issues that are already well developed in policy documents. However, the NSSD fails to account for new emerging issues. Therefore, the situation of sustainable development in Lithuania is biased as well as poorly updated with new issues, risks, opportunities and challenges. The strategy does not mention issues related to the social dimension of the broader sustainable development paradigm, such as the declining or inadequate quality of public social services, problems of emigration reaching close to 20% of the labor force in 2006 and a new wave of labor force immigration (CSI 2005). Similarly, ecological concerns are poorly addressed where they should have been prioritized, such as with regard to efficient use of natural resources, forestry sustainability and the impact of urban sprawl, especially in natural parks (Mžavanadze 2008). Additionally, economic costs of applying sustainability concerns were not clarified with regard to the energy security of Lithuania. This is in the light of the approaching closure of the Ignalina nuclear power plant and the very sparse application and use of renewable energy resources.

The sectoral SWOT analysis is followed by short-term, mid-term and long-term goals and actions to achieve them. Most of them stem from the EU *Acquis* obligations for Lithuania, which are quite concrete and well elaborated. As previously mentioned, the goals and the prescribed means of achieving them include only governmental responsibilities, excluding businesses, civil society and the individual.

Despite the disadvantages of the NSSD, it succeeds in representing a certain vision of sustainable development. The problem is that this vision is limited to that of state bureaucrats and academics. It also explains why the strategy has been professionally elaborated. There were fewer interests to reconcile, otherwise the process would have taken much longer and there would have been a greater degree of vagueness in the contents. It appears that today the Lithuanian NSSD is a bureaucratic vision of sustainable development, leaving aside business, civil society and the individual. At the same time it creates a false image that sustainable development is something only governments and public administrations are concerned with and the rest of society has no responsibility over.

Impact on policy implementation: no implementation

Vertical policy integration. 'The concept of sustainable development transcends not only the competences of ministries or departments within a government, but also different levels of jurisdictions, from the European Commission to city halls' according to Steurer and Martinuzzi (2005, p. 462). A survey of NSSDs in Europe concludes that vertical integration is either weak or non-existent (Steurer & Martinuzzi 2005).

The Lithuanian NSSD rarely mentions local governments. According to the document, the municipalities are leaders and partners in the implementation of the NSSD (NSSDoL 2003, p. 8). The biggest issue concerning municipalities seems to be the unequal economic development of regions (NSSDoL 2003, p. 37). A lot of hope is invested in EU structural funds, which are supposed to resolve developmental inequality. However, Lithuania is treated as an undivided region for EU structural funds; therefore, there is a risk of further increase in regional differences, with the funds flowing to the centers. All municipalities of Lithuania are expected to adopt Local variants of Agenda 21 (NSSDoL 2003, p. 41) and the documents should be promoted at the state level (NSSDoL 2003, p. 84), with the state providing financial and methodological support for their preparation and implementation (NSSDoL 2003, p. 84)

The implementation of vertical integration has not been consolidated with the necessary human and financial resources or technical expertise from the state level. The strategy requires the municipalities to transpose the principles elaborated in the NSSD into municipal development plans. However, at the time of writing, no monitoring and control of this process have been initiated. There is no official support or pressure for municipalities to adopt Local Agenda 21. The *de facto* implementation of Local Agenda 21 and the NSSD of Lithuania is voluntary for the municipalities. No resources and incentives are being provided by the central government to municipalities to include sustainable development principles in their activities.[1] At the time of writing, only 30% of Lithuanian municipalities participate in Local Agenda 21 initiatives and projects and only 10% seek to approve them politically and incorporate sustainable development principles into their strategic planning activities.[2]

Failure to involve local governments in strategy formation is another obstacle to successful collaboration in this field. The exclusion of local authorities translates into a lack of ownership and commitment for the initiative. There has been an idea to create a separate forum on sustainable development for local affairs – another NSSD implementing body – but it has never received the necessary political support at the state level.

The biggest obstacle to sustainable development practices at the level of the municipalities is the fast rotation of local politicians and their negative attitude towards any kind of training. Every generation of politicians has to be educated and trained on sustainable development issues at the beginning of every new term. Therefore, the focus is now on sustainability education and training of the public administration staff rather than the politicians. This tendency confirms Steurer and Martinuzzi's (2005) dichotomy of horizontal integration at municipal level emerging between public administrators and politicians. The administrators tend to take the lead and politicians allow them to proceed with limited scope and are probably not interested in the specific formulations of strategic documents.

Finally, it seems that Lithuania's administrative system needs to be reformed. The main problems of the current administrative and local government structures include high centralization of decision-making, low stakeholder involvement and bureaucratic obstacles for bottom-up societal initiatives. The suggested reforms include positive changes for the governance of sustainable development: a change to the election system, the elimination of regional counties (an unnecessary level, and a relic of the Soviet government system), eliminating disproportions among local governments and creating more municipalities (municipalities in Lithuania are among the largest in Europe (Bėkšta & Petkevičius 2000)), strengthening of the local counties (the smallest and lowest administrative units), eliminating barriers to public participation and bottom-up approaches to the multi-level public administration system (Bakučionis 2006). One positive sign in recent years is that there has been a growing number of active local communities raising issues of public interest (Vilnius old town conservation community, Pilaitė community, the local communities movement *Piliečių santalka*, public spaces' community, etc.); however, significant barriers exist to their bottom-up initiatives (Žiliūkaitė *et al.* 2006). All of the suggested reforms are in line with the subsidiarity and public participation principles of sustainable development.

Horizontal policy integration. The pursuit of horizontal policy integration in the field of environment and sustainable development began with the Bruntland report (WCED 1987; Steurer & Martinuzzi 2005). It aims to change the perception that environment (and sustainable development) is a narrow expertise issue. In order for environmental policy to achieve its goals, there is a need to integrate environmental objectives into every policy-making field, to remove contradictions and barriers and focus on achieving common goals.

The Lithuanian NSSD states that 'formulation of priorities and objectives of the Strategy followed national interests and already prepared strategic documents' (NSSDoL 2003, p. 4). From the beginning, the NSSD of Lithuania became a subordinate strategy to the main strategic documents in Lithuania and does not aim to challenge them. However, another unofficial source states that 'the Strategy for Sustainable Development of the State is at the top of the pyramid of the different planning documents and serves as the basis for development of sectoral strategies for three to seven years. And vice versa – sectoral strategies can lead to changes in the Strategy for Sustainable Development of the State' (Rekerta 2004, p. 2). However, this is not an official statement and it is the second relationship that is more prevalent in the case of the Lithuanian NSSD. A study by the World Bank on public-sector capacities shows positive progress in strategic management and planning in public policy (World Bank 2006). From a practitioner's point of view the processes of strategic management and policy-making in the public administration lack coordination between them and are considered to be erratic (Leichteris 2006).

According to Steurer and Martinuzzi (2005) one of the biggest positive outcomes from the NSSD processes is the realized cross-sectoral linkages, but in the case of Lithuania it is an unlikely outcome. Academics have led the NSSD formulation and review process and there is no sense of ownership among policy-makers. The document simply summarizes sectoral actions, which fit under the umbrella of sustainable development. It is important for the NSSDs to build on the existing national planning and policy-making processes. However, the NSSD should be more than a summary; it should offer new initiatives. Therefore, it is safe to claim that this mutual learning between sectoral ministries has not happened. Nevertheless, the strategy mentions that its provisions must be integrated into other relevant strategic documents, but it does not suggest any monitoring.

According to Steurer and Martinuzzi (2005), it is important to have horizontal integration mechanisms at both levels – political and administrative. The Lithuanian NSSD process foresees only one horizontal integration mechanism – the National Commission on Sustainable Development. It embraces horizontal integration at both administrative and political levels, since the majority of its members are government cabinet members and representatives of different ministries. The commission meets once or twice a year when NSSD review or monitoring processes are initiated. In addition, the variety of professionals involved is probably too wide to discuss concrete practical measures.

Horizontal policy integration is only possible with strong political support and commitment. Despite the presence of high-level political figures, the political commitment to the sustainable development paradigm in policy-making starts and ends in those rare meetings of the commission. Today in Lithuania there is no strong political leadership for the advancement of sustainable development issues. Neither the politicians nor the public officials perceive sustainable development as a useful paradigm for policy-making. A perception exists that Lithuania has already been developed in a sustainable way and does not require further improvement.[3]

Policy implementation and monitoring. Policy implementation and monitoring are the final stages of the NSSD policy cycle, which generate important feedback for the new policy cycle that enables institutional learning. Therefore it is important to explore these last two stages in order to estimate the capacity for institutional learning and the success of NSSDs as continuous and self-improving social initiatives.

Separation of the formulation and implementation stages is considered to be one of the biggest failures in strategic management according to Mintzberg *et al.* (1998). The Lithuanian NSSD presents an interesting case of what is in effect a triple dichotomy in terms of policy implementation. Strategy formulation has been outsourced to academics, who have summarized the administration's plans to fit under the sustainable development umbrella. Politicians are uninterested. Therefore, the implementation of the NSSD becomes an easy thing – it is being implemented without any special efforts and it does not require any resources. There are neither human nor financial resources allocated for implementation.

Monitoring of the NSSD is narrowly understood as a comprehensive analysis of specified quantitative indicators, without any qualitative insights into integration of sustainable development principles, issues in policy-making, budgetary processes, internal or external evaluations or peer reviews. Therefore, the understanding of sustainable development implementation and monitoring in Lithuania is reduced to the observation of performance of the national eco-efficiency indicators (Juknys 2006; NSSDoL 2003).

Discussion

This essay has sought to explore the potential uses, benefits and conditions of social capital involvement in policy-making and has been built on a case study of Lithuanian sustainable development policy. This case study is an illustrative example of how novel policies based on advanced governance norms and traditions are being implemented in societies that do not engage in participative formulating practices. Some studies of policy-making have previously suggested that policies regarding sustainable development are devised because of institutional pressure, or due to international conditioning. This investigation indicates that the Lithuanian top-down approach to NSSDs has been a result of a limited participatory ethos within the population and was due to differences in norms and perceptions on various levels of governance and within Lithuanian society. For these reasons, the development of necessary social capital for sustainable policy development and implementation has never gained recognition or support.

From a comparative perspective, the Lithuanian NSSD experience has proved similar to that of other member states of the European Union. For instance, most NSSD processes have received very limited public participation. Many NSSDs are vague and in some cases crucial mechanisms, such as horizontal and vertical integration, have been omitted (Steurer & Martinuzzi 2005). Furthermore, capacities and resources have rarely been sufficient for successful realization of NSSDs. However, some case studies can provide fragments of good-practice examples. Slovakia is a renowned case study for effective and wide public participation. Germany and the United Kingdom could serve as models of horizontal policy integration mechanisms. Switzerland, with a Sustainable Development Forum for its towns and cantons, could serve as a model for vertical policy integration and Austria could be distinguished by virtue of its complex and flexible mechanisms of policy-making and coordination (Steurer & Martinuzzi 2005).

Several lessons can be learned from Lithuania's experience. On the one hand, the failure to draft policy in cooperation with members of society and an expectation of implementing it effectively indicates strong legacies of the socialist technocratic relations between the state apparatus and the society affected. Although Lithuania has joined the EU and thus conforms to the bulk of European legislation, it is difficult to see how certain legislative measures are transposed into policy action and how they are implemented on the ground. However, the Lithuanian NSSD has not achieved the role assigned to it by the policy-makers. This study points out the limited effect that European conditioning has had on its acceding members. While the EU *Acquis* and legislation have been incorporated and transposed into local laws and regulations, these have barely affected public engagement in policy-making.

This point, however, should not lead one to believe that limitations of strategic planning and management are embedded in the Lithuanian NSSD. It is a professionally written strategic document following the principles of the planning school (Mintzberg *et al.* 1998). Its positive feature is that it is concise and concrete, since many other NSSDs have been too vague (Steurer & Martinuzzi 2005). Nevertheless, the Lithuanian NSSD also testifies to the negative inheritance of the planning school: 'strategic planning has never been strategy making' (Mintzberg *et al.* 1998, p. 77). The strategy-making process is hierarchical with a closed and narrow formulation circle relying on professional expertise, aiming at forecasting the future situation and adapting to it with quantified goals and means of control. The Lithuanian NSSD is a plan, but not a strategy, because it does not aim to challenge the current situation; it rather adapts to it and flows with the current of time.

From the public policy perspective, Lithuanian policy-making has not yet made a transition from the strong government paradigm to a governance paradigm. The sustainable development policy formulation process has been closed, involving only a narrow circle of public administration and scientific expertise, while the contents only consider the responsibilities and goals of the government. The policy-makers fail to build on softer steering mechanisms and fail to create a stakeholder forum for such a multifaceted issue as sustainable development. This approach results in low awareness and participation, and poor implementation of the policy. Initially intended to become a novel policy-making process, encompassing multi-stakeholder dialog and involving the public, the sustainable development policy ended up back on the regular track of policy-making in Lithuania.

From the perspective of civil society and social capital, governance for sustainable development should build on multi-stakeholder processes that have the potential to form new partnerships, change behavior and beliefs, generate commitment and trust, spread relevant information and foster innovation for sustainable development. Failure to build on existing social capital and to invest in the additional creation of social capital also leads to a failure of the sustainable development initiative, as has happened in Lithuania. Instead of becoming a platform for social cohesion and consolidated long-term effort for sustainable development, the NSSD has become an ineffective policy unable to deliver tangible results. The state fails to comprehend the crucial role of social capital and civil society in public policy, while Lithuanian society also suffers from a severe disability syndrome – disbelief that the collective action of citizens can make a difference.

Neither political nor administrative ownership of the NSSD processes ever existed – the NSSD processes have been outsourced to academia. This has resulted in a contrast between academic, professional expertise and general public knowledge. As a consequence, one of the crucial elements of sustainable development processes (public participation) has been eliminated. Instead of a societal consensus document, bridging the policy-makers and the empowered civil society, Lithuania only has a strategic bureaucratic plan with prevailing 'sectoralism' and no tangible improvement in policy-making practices as a result of this initiative. Civil society inputs have been replaced by stronger governmental control, which in turn further accelerates the vicious cycle of prejudices of strong omnipotent government and the undiscovered, unprofessional civil society.

It remains to be seen how the NSSD will be implemented and coordinated in the coming years. Some of the outlined weaknesses of the policy document can be effectively overcome if adequate human and financial capacities for its implementation are allocated by the government, thus providing for an increasing exchange between the stakeholders both vertically and horizontally. Additional funds for implementation of the existing program and the development of its objectives with input from the civil society organizations and local people would further facilitate monitoring of subsequent drafting measures and development of addenda to the existing agenda. Should these changes find support with the policy makers, the 'ownership' of the strategy formulation is likely to shift from the administrative and political level to members of Lithuania's civil society. This will undoubtedly ensure that future NSSD programs have greater relevance for the society they affect, and also establish a tight connection between the individual endowment with social capital and policy initiatives building upon them.

Acknowledgements

I would like to thank Aleh Cherp (Central European University, Hungary) for the primary inspiration to work on this issue, Romualdas Juknys (Vytautas Magnus

University, Lithuania) for his kind invitation to be a part of the Review Panel of National Strategy for Sustainable Development of Lithuania, and Renata Dagiliūtė (Vytautas Magnus University, Lithuania) and Alessandro Bonifazi (Polytechnic University of Bari, Italy) for insightful comments. I would like to extend my thanks to David Galbreath (University of Aberdeen, UK) for his invitation to publish in the *Journal of Baltic Studies*, and to Timofey Agarin (University of Aberdeen, UK) and the two anonymous peer-reviewers for help and suggestions. I am grateful to Julia Leventon (Central European University, Hungary) and Matt Bond for proofreading. All the interpretations in the essay are my own and I bear the sole responsibility for mistakes.

Notes

1 Environmental Centre For Administration and Technology (ECAT), *Appeal to the state institutions, the President, the Parliament and the Government of Lithuania signed by the participants of the conference 'Contemporary aspects of the implementation of local sustainability'* organized by the Environmental Centre for Administration and Technology, 30–31 October 2003.

2 Interview with A. Alijošiūtė, director of Environmental Centre for Administration and Technology (ECAT), 18 July 2007, Kaunas, Lithuania.

3 *Žinių Radijas* [News Radio of Lithuania], *Antradienio 'Raktas'. Darnus vystymasis: ar strategija virsta praktika? [Tuesday's broadcast 'Key'. Sustainable Development: does the strategy become practice?]*. With participation of R. Pakalnis, co-author of the NSSD of Lithuania and member of the Review Panel of NSSD of Lithuania. Broadcast on 5 June 2007, available at: http://www.ziniur.lt/apie/pranesimai?item=1219, accessed 16 January 2009.

References

Adam, F. & Roncevic, B. (2003) 'Social Capital: Recent Debates and Research Trends', *Social Science Information*, 42, 2, pp. 155–83.

Adler, P. S. & Kwon, S.-W. (2000) 'Social Capital: The Good, the Bad, and the Ugly', in Lesser, E.L. (ed.) (2000) *Knowledge and Social Capital* (Woburn, Butterworth-Heinemann).

Agenda 21 (1992) United Nations Conference on Environment and Development (UNCED), Rio de Janerio, Brazil, 3–14 June, available at: http://www.un.org/esa/sustdev/documents/agenda21/english/agenda21toc.htm, accessed 13 January 2009.

Baker, S. (2000) 'The European Union: Integration, Competition, Growth—and Sustainability', in Lafferty, W. M. & Meadowcroft, J. (eds) (2000) *Implementing Sustainable Development: Strategies and Initiatives in High Consumption Societies* (Oxford, Oxford University Press).

Bakučionis, T. (2006) *Savivaldos reforma – Seimo įkaitė? [Local government reform – hostage of the parliament?]*, available at: http://www.bendruomenes.lt/news.php?id=38058&strid=11791, accessed 26 July 2007.

Bebbington, A., Guggenheim, S., Olson, E. & Woolcock, M. (2004) 'Exploring Social Capital Debates at the World Bank', *Journal of Development Studies*, 40, 5, pp. 33–64.

Bėkšta, A. & Petkevičius, A. (2000) 'Local Government in Lithuania', in Horvath, T. (ed.) (2000) *Decentralization: Experiments and Reforms*, Vol. 1 (Budapest, Open Society Institute), available at: http://lgi.osi.hu/publications/2000/25/Chapter_4.PDF, accessed 16 January 2009.

Cherp, A., George, C. & Kirkpatrick, C. (2004) 'A Methodology for Assessing National Sustainable Development Strategies', *Environment and Planning C: Government and Policy*, 22, 6, pp. 913–26.

Civil Society Institute (CSI) (2005) *Lietuvių emigracija: problema ir galimi sprendimo būdai* [*Lithuanian emigration: a problem and possible solutions*], available at: http://www.civitas.lt/files/Tyrimas_Lietuviu_emigracija_Studija.pdf, accessed 16 January 2009.

Coleman, J. S. (1988) 'Social Capital in the Creation of Human Capital', *The American Journal of Sociology*, 94, S1, pp. S95–S120.

Devine-Wright, P., Fleming, P. D. & Chadwick, H. (2001) 'Socio-psychological Perspective: Role of Social Capital in Advancing Regional Sustainable Development', *Impact Assessment and Project Appraisal*, 19, 2, pp. 161–7.

Doyle, T. (1998) 'Sustainable Development and Agenda 21: the Secular Bible of the Global Free Markets and Pluralist Democracy', *Third World Quarterly*, 19, 4, pp. 771–86.

EUCOM (Commission of the European Communities) (2005a) *Draft Declaration on Guiding Principles for Sustainable Development*, Communication from the Commission to the Council and the European Parliament, COM(2005) 218 final. Brussels, 25 May, available at: http://eur-lex.europa.eu/LexUriServ/site/en/com/2005/com2005_0218en01.pdf, accessed 11 December 2008.

EUCOM (Commission of the European Communities) (2005b) *The 2005 Review of the EU Sustainable Development Strategy: Initial Stocktaking and Future Orientations*, Communication from the Commission to the Council and the European Parliament, COM(2005) 37 final, Brussels, 9 February, available at: http://eur-lex.europa.eu/LexUriServ/site/en/com/2005/com2005_0037en01.pdf, accessed 11 December 2008.

Harriss, J. (2002) *Depoliticizing Development: The World Bank and Social Capital* (London, Anthem Press).

Institute for European Environmental Policy (IEEP) (2006) *Prime-SD. Peer Review Improvement through Mutual Exchange on Sustainable Development. A guidebook for peer reviews of national sustainable development strategies*, Guidebook prepared for DG Environment, available at: http://ec.europa.eu/environment/pdf/nsds.pdf, accessed 21 January 2008.

Islam, M. K., Merlo, J., Kawachi, I., Lindstrom, M. & Gerdtham, U. G. (2006) 'Social Capital and Health: Does Egalitarianism Matter? A Literature Review', *International Journal for Equity in Health*, 5, 3.

Juknys, R. (2006) 'Implementation of National Strategy for Sustainable Development: Achievements and Problems', in *Vilnius, September 28–30, 2006. Citizens and Governance for Sustainable Development* (Vilnius, VGTU Press 'Technika').

Kekic, L. (2007) *The Economist Intelligence Unit's Index of Democracy*, available at: http://www.economist.com/media/pdf/DEMOCRACY_INDEX_2007_v3.pdf, accessed 13 January 2009.

Lafferty, W. M. (2004) 'Introduction: Form and Function in Governance for Sustainable Development', in Lafferty, W. M. (ed.) (2004) *Governance for Sustainable Development: The Challenge of Adapting Form to Function* (Cheltenham, Edward Elgar).

Lehtonen, M. (2004) 'The Environmental–Social Interface of Sustainable Development: Capabilities, Social Capital, Institutions', *Ecological Economics*, 49, 2, pp. 199–214.

Leichteris, E. (2006) *Verslo konkurencingumo didinimas – ar turime kompleksinį požiūrį? [Increasing Business Competitiveness—do we have a complex view?]*, Presentation delivered at the conference 'Competitiveness of Lithuania in the global economy: challenges, dangers and opportunities. Case studies of Slovenia and Lithuania', Vilnius, Lithuania, 21 September 2006, available at: http://www.klasteriai.lt/bylos/eleichteris.ppt, accessed 13 February 2008.

Lele, S. (1991) 'Sustainable Development: a Critical Review', *World Development*, 19, 6, pp. 607–21.

Mačiulytė, J. & Ragauskas, P. (2007) *Lietuvos savivalda: savarankiškos visuomenės link? [Lithuanian Self-Government: The Path to a More Civil-Minded Community?]* (Vilnius, Aureus Versus).

Meadowcroft, J. (2004) 'Participation and Sustainable Development: Modes of Citizen, Community and Organizational Involvement', in Lafferty, W. M. (ed.) (2004) *Governance for Sustainable Development: The Challenge of Adapting Form to Function* (Cheltenham, Edward Elgar).

Mihaylova, D. (2004) *Social Capital in Central and Eastern Europe. A Critical Assessment and Literature Review* (Budapest, Central European University, Center for Policy Studies).

Mintzberg, H., Ahlstrand, B. W. & Lampel, J. (1998) *Strategy Safari: The Complete Guide Tour Through the Wilds of Strategic Management* (London, Pretince Hall).

MoEoL (Ministry of Environment of Lithuania) (2002) *Subalansuotos plėtros įgyvendinimo nacionalinė ataskaita. Nuo Rio de Žaneiro link Johanesburgo. Nuo pereinamojo laikotarpio link subalansuotos plėtros [National report on sustainable development. From Rio de Janeiro to Johannesburg. From transition to sustainable development]*, available at: http://www.am.lt/LSP/files/SUB-PLET-NAC-AT.pdf, accessed 14 January 2009.

MoEoL (Ministry of Environment of Lithuania) (2005) *Nacionalinės darnaus vystymosi strategijos įgyvendinimo 2003–2004 metais ataskaita [National report on implementation of sustainable development strategy in 2003–2004]*, available at: http://www.am.lt/VI/files/0.5122070011586614 37.pdf, accessed 14 January 2009.

Mžavanadze, N. (2008) *Sustainability in Transition: A Case Study of Lithuania in 1990–2004* (Saarbrucken, VDM Verlag).

NSSDoL (2003) National Strategy For Sustainable Development of the Republic of Lithuania, Approved by resolution No 1160 of 11 September 2003 of the Government of the Republic of Lithuania, available at: http://www.am.lt/VI/en/VI/files/0.901665001073997792.pdf, accessed 16 January 2009.

Organization for Economic Co-Operation and Development (OECD) (2001a) *Citizens as Partners: Information, Consultation and Public Participation in Policy-making* (Paris, OECD).

Organization for Economic Co-Operation and Development (OECD) (2001b) *The DAC Guidelines. Strategies for Sustainable Development: Guidance for Development Co-operation*, available at: http://www.oecd.org/dataoecd/34/10/2669958.pdf, accessed 2 June 2008.

Ostrom, E. (2000) 'Social Capital: A Fad or a Fundamental Concept', in Dasgupta, P. & Serageldin, I. (eds) (2000) *Social Capital: A Multifaceted Perspective* (Washington, World Bank Publications).

Pretty, J. & Ward, H. (2001) 'Social Capital and the Environment', *World Development*, 29, 2, pp. 209–27.

Rekerta, K. (2004) 'Lithuania: Strategic Planning—Seeking Effectiveness', *Sigma Update. Support for Improvement in Governance and Management*, Issue No. 2 on Accountability Structures. A joint initiative of the OECD and the European Union, available at: http://unpan1.un.org/intradoc/groups/public/documents/NISPAcee/UNPAN014625.pdf, accessed 13 February 2008.

Schuller, T. & Field, J. (1998) 'Social Capital, Human Capital and the Learning Society', *International Journal of Lifelong Education*, 17, 4, pp. 226–35.

Spangenberg, J. H. (2006) 'Towards a European methodology for NSDS reviews', in *International Conference on Sustainability Measurement and Modelling (ICSMM 06)*, Barcelona, November 2006.

Steurer, R. & Martinuzzi, A. (2005) 'Towards a New Pattern of Strategy Formation in the Public Sector: First Experiences with National Strategies for Sustainable Development in Europe', *Environment and Planning C: Government and Policy*, 23, 3, pp. 455–72.

United Nations Department of Economic and Social Affairs (UNDESA) (2002) *Guidance in preparing a National Sustainable Development Strategy: managing sustainable development in the new millennium*, available at: http://www.un.org/esa/sustdev/publications/nsds_guidance.pdf, accessed 21 January 2008.

United Nations Development Programme (UNDP) (2005) *JTVP ir Lietuvos partnerystės apžvalga (1992–2005m.)* [*Overview of the partnership between UNDP and Lithuania in 1992–2005*] (Vilnius, UNDP).

VanDeveer, S. D. & Carmin, J. (2005) 'Assessing Conventional Wisdom: Environmental Challenges and Opportunities beyond Eastern Accession', in Carmin, J. & VanDeveer, S. D. (eds) (2005) *EU Enlargement and the Environment: Institutional Change and Environmental Policy in Central and Eastern Europe* (London, Routledge).

Woolcock, M. (2001) 'The Place of Social Capital in Understanding Social and Economic Outcomes', in *The contribution of human and social capital to sustained economic growth and well-being*, OECD international symposium report, pp. 65–88, available at: http://142.236.154.1/sp-ps/arb-dgra/publications/books/oecd/en/5-woolcock.pdf, accessed 27 May 2008.

World Bank (2006) *Public Sector Capacity in the EU 8. Report on Strategic Planning and Policy Management In Lithuania and Latvia*. Background paper, October, available at: http://www.lrv.lt/strateginis/Pasaulio%20banko%20studija.pdf, accessed 14 June 2008.

World Commission on Environment and Development (WCED) (1987) *Our Common Future* (the reference is to the 1990 edition, Oxford University Press, Oxford).

Wright, P., Pringle, C. & Kroll, M. (1992) *Strategic Management Text and Cases* (Needham Heights, MA, Allyn and Bacon).

Žiliūkaitė, R., Ramonaitė, A., Nevinskaitė, L., Beresnevičiutė, V. & Vinogradaitė, I. (2006) *Neatrasta galia: Lietuvos pilietinės visuomenės žemėlapis* [*Undiscovered Power: Map of the Civil Society in Lithuania*] (Vilnius, Versus aureus).

NARVA–IVANGOROD: INTEGRATING AND DISINTEGRATING TRANSBOUNDARY WATER NETWORKS AND INFRASTRUCTURE

Jussi S. Jauhiainen and Tarmo Pikner

Introduction

Water is a necessary everyday commodity, and every modern society is dependent on urbanized water. Nevertheless, people take water for granted. People do not think about drinking or sewage water networks, because water simply comes and goes. However, a hybrid and heterogeneous network involving water intertwines nature and human activities. This socionature varies with regard to its stability, extension and forms of time and space. A key scholar of the geography of water, Karen Bakker (2005, p. 546), has stated that the analysis (of water networks and systems) should focus on specific socionatures as active subjects of neoliberalization in specific historical–geographical contexts.

In this essay, water networks are contextualized in the transboundary area of Narva–Ivangorod. The aim is to reveal how socionatural networks involving water emerge and the kind of commodity water has become. The study area, which consists of around 80,000 inhabitants, is located on the border of Estonia and Russia on two sides of the Narva River. Throughout history, Narva and Ivangorod have been united and separated. Today, Narva belongs to Estonia while Ivangorod remains in Russia.

This essay analyses water networks and infrastructure in Narva–Ivangorod from the capitalism of the early nineteenth century through the state socialism of the twentieth century to the current post-socialist era. During the period in question, the study area has been part of the Russian Empire, independent Estonia, and the Soviet Union. Now the area again belongs to independent Estonia, which is part of the European Union. The main focus of analysis here is the period following the division of the area between Estonia and Russia during the early 1990s. The transformation and disruption of water infrastructure and networks are studied in a transboundary context. The organization of this socionature involves environmental governance (see Davidson & Frickel 2004). The case study was conducted from three integrated viewpoints. The first deals with urbanization of water and its circuits (Kaïka 2005; Swyngedouw & Kaïka 2000) through changing water-related infrastructure (Graham & Marvin 2001). The second considers the formation of water as a commodity (Appadurai 1986; Bakker 2005; Davis 2005). The third looks at the politics of transforming water networks.

To analyze the case material methodological triangulation is used. Consideration is given to files from the historical archives of Narva and its water authorities, which describe events and processes of transboundary water infrastructure and related practices from the late nineteenth century onwards. This material also includes recent business reports by the local water company Narva Vesi. Nine themed interviews were also conducted in January 2005 with key persons involved in water infrastructure and delivery in Narva and Ivangorod. The focus was on the post-Soviet period. The media analysis focused on *Postimees*, the largest daily newspaper in Estonia, which during 1996–2007 published a total of 42 articles on the water issue in Narva–Ivangorod in 1996–2007. Qualitative content analysis of the articles and interviews focused on the functioning of and changes in the water networks, management of water across the border, arguments to modify drinking water and sewage flows, and the renovation of water infrastructure. In addition, several visits and the authors' fieldwork in Narva and Ivangorod over the past ten years has helped to contextualize the study material.

Water-related issues in a transboundary context can be approached from a variety of standpoints, including ecology, economics, sociology, politics and planning. In addition, agency capture can be used to analyze the formation of water networks (Davidson & Frickel 2004, p. 474). This essay focuses on the commodification of water in the particular cross-border context of Eastern Europe and underscores the importance of neoliberal politico-economic tendencies to any discussion of this phenomenon. An analysis of the neoliberal commodification of water, we argue, benefits from a recognition of extended social agency, including hybrid interactions between human and nonhuman entities (see Castree & MacMillan 2001; Whatmore 2005). Neoliberal commodification associates water with different resources as valuable things. Through this the positions of social agency and wider socio-natural networks are also transformed. This leads to an understanding that a single (collective) actor in the form of a water network is actually a sum of many hybrid actors, including (natural) non-human entities and humans. Besides the spatial dimension of water networks, one needs to consider particular factors and politics that are important in the formation of infrastructure links and commodity values.

The water network, including commodity flows, evolves as a consequence of relations between entities including, for example, material resources, managers and technical artifacts. The analysis of environmental governance needs to discuss critically the particular relations that change interactions between nature and society. Circulations of water are particularly important for understanding the modern process of urbanization that transforms natural water into a resource with various (environmental) values (see Kaïka 2005). The neoliberal politics of water commodification is discussed within the glocal context, including the governance of infrastructure links, multiple values of material things and everyday social practices. This essay takes into account similar transformations of water in Europe (see Bakker 2005; Bressers & O'Toole 1994) and elsewhere (see Budds & McGranahan 2003; Davis 2005).

Water in networks

Water is a resource with particular biophysical, sociocultural and spatial characteristics. It both carries and is part of other socionatural flows. Water networks bring heterogeneous places, people, buildings and urban elements into a dynamic relationship. The topology of water networks moves seamlessly from local to global and from non-human to human. Distribution grids and topological connections link systems and practices of production and consumption. Consequently, water networks connect and disconnect places and configure social relations in these places (Graham & Marvin 2001, pp. 10–11).

A water network is always socially complex, spatially continuous and hybrid. A number of economic, political, social and ecological issues are subsumed within water networks, such as basic consumption needs, trade, management, safety, etc. in water and the usage of water infrastructure (Finger *et al.* 2006, pp. 12–32). Swyngedouw (1999) suggests that water embodies multiple tales of socionature as 'hybrid':

> The cyborg flow of H_2O relates all things/subjects in a network, a rhizome, connecting the most intimate of socio-spatial relations, inserts them in a political-economy of urban, national and international development and is part of a chain of local, regional, national and global circulations of water, money, texts and bodies. (Swyngedouw 2004, p. 120)

Studying networks more closely, one sees how water is transformed into a commodity or 'thing' – an economic good with an exchange value. A conventional political economy perspective underlines the importance of market incentives and private companies, which link water networks, on the one hand, to theories of ecological Marxism and, on the other, to ecological modernization (Davidson & Frickel 2004, pp. 475–7). The measurement of technical and economic efficiency of water supply systems becomes significant when water provision becomes a business activity (Bakker 2001, p. 144). This commodification (i.e. the substantial role of pricing, measuring, distributing, etc. of water) adds new qualities and artifacts to socionatural water networks. In fact, the transformation of a once free, and later state-managed, common resource, such as water, into a measured, monetized, privatized and salable good is an ordinary phenomenon in Western societies (Castree 2003). Increasingly, water is produced, distributed and exchanged primarily according to the economic profit imperative (Bakker 2001; Davis 2005; Smith 1984). However, such a transformation from the value of water use into exchange value often involves contradictions and conflicts. According to Kallis & de Groot (2003, pp. 224–5) and Bithas (2008, p. 1293), these European Union water policy measures aim to implement more efficient and sustainable water management models that can take into account market changes and also cope with decreasing suitable water resources in European urban areas and their comprehensive management at different spatial scales (Finger *et al.* 2006, pp. 172–93). The privatization of water delivery systems was actively taking place in the 1990s in Africa, Asia and Latin America (see Budds & McGranahan 2003).

When discussing water supply, Bakker (2005) makes a distinction between the privatization, commercialization and commodification of water. *Commodification* means a process whereby goods formerly outside market spheres of existence enter the world of money. Water becomes exchangeable and sellable through the medium of money. This transformation of biophysical objects, such as water, takes place simultaneously in three spheres: discursive, material and socio-economic. Water is abstracted, valued and displaced in the discursive sphere. This means that water is separated from its ecological context. In this case, it means the separation of drinking water from the river by physical interventions and adaptations in the material sphere. In addition, price, charge, allocation, and exchange methods as regards water are introduced in the socio-economic sphere. *Commercialization* requires the implementation of management practices supporting commercial principles (efficiency), methods (cost-benefit analysis), and objectives (profit-maximization). *Privatization* means a change from public to private ownership in the organization and practices of water supply. However, Bakker (2005) rightly observes that commodification is a contested, partial and transient process in which objects shift back and forth between commodity and non-commodity status.

Commodification often takes place through various events and socio-technical mediations, events and ruptures. The transformation of water is not necessarily a linear process. In fact, for many, the infrastructure becomes visible only when something 'goes wrong' within infrastructure networks (Law 2000) and the matter passing through the networks does not reach the destination as agreed. Such failures may rest on a combination of factors, ranging from material defects to politics. The following case study of Narva–Ivangorod water networks discusses socio-technical ruptures as part of a wider analysis of the neoliberal commodification of water within a global transboundary context.

The case of Narva–Ivangorod

The transformations of transboundary water links and drinking water values between Estonia and Russia are illustrative of a larger shift from the Soviet economic system to Western capitalism, including privatization and new values for things as commodities. However, the changes in drinking and sewage water circulations between Narva and Ivangorod can only be fully understood through a contextualized case study that elaborates upon social agency and spatiality (see Whatmore 2005) and discusses local water networks in relation to cases in other countries.

The obvious starting point for this case study is the Narva River; this brings together non-human and human actors who construct the water network materially, socially and culturally (Kortelainen 1999, p. 237; see also Eden *et al.* 2000).

The complex and hybrid socionatural water network around the Narva River has emerged and continues to evolve through historical–geographical processes, which individualize water and transform its character in an urban and political context. The urbanization of nature through the transformation of wild water into urban water weaves together the urban fabric and ecological, economic, social and political networks that metabolize nature (Swyngedouw & Kaïka 2000, p. 569). According to Kaïka (2005), the urbanization of nature includes, besides commodity formations, symbolic aspects and uses of water such as public fountains in cities. This case study cannot analyze all such topics, but recognizes the variety of symbolic and everyday uses of water in Narva–Ivangorod throughout history. The focus is on the commodification of water resources in the historical-geographical context.

Integration of water infrastructure networks

Narva, a stronghold under the Danish kingdom, was established in 1256 on the banks of a river slightly inland from the Baltic Sea. Such a location was important because of the trade between central European regions and Russia and its predecessors. The Vikings had earlier used the same river for their forays into what is today Russia. The area has been inhabited for a very long time. In 1345, Narva received town rights from the King of Denmark. The small town soon passed to the Teutonic order and then to the Livonian knights, who possessed it until the mid-sixteenth century. During the Livonian era, the Hermann Fortress was built next to the river (Eesti Entsüklopeedia 2003). Narva gradually grew in importance, and in 1492 the Grand Prince of Muscovy Ivan III founded a stronghold on the other side of the river to facilitate the Russians' access to the Baltic Sea and to serve as protection against expanding western forces. Next to the stronghold grew a settlement that became known as Ivangorod, 'the town of Ivan'. Occasionally there were hostilities over the river that was a dividing element between the two settlements. In fact, troops from the Russian side destroyed Narva several times.

In 1558, Narva and Ivangorod became a unified urban area of 5,000–7,000 inhabitants after Russian troops conquered Narva. Narva–Ivangorod became more important and competed in trade with the Hanseatic town of Tallinn. Sweden conquered the area in the latter half of the sixteenth century, and in the following century both the stronghold and the town developed in the Baroque style. At the beginning of the eighteenth century, the Russians conquered Narva–Ivangorod, which lost its significance as a border town and was instead developed as a harbor for the transportation of wood (Eesti Entsüklopeedia 1994b). Water in the form of the river was fundamental to the development of the settlement because of trade and because the river regulated the traffic between Narva and Ivangorod. Obviously, water from the river was also used in everyday chores.

The nineteenth century was the period of modernization of Narva–Ivangorod. In 1819, the first textile factory was established; a few years later, it utilized the first steam-power machine in Estonia. The town of a few thousand inhabitants started to grow when new textile factories were established. Kreenholm was founded in 1857, and later it became the largest textile factory in Europe. The factory brought various water-related innovations to Estonia, including several machines that utilized waterpower (Eesti Entsüklopeedia 1994a).

In addition, in the later nineteenth century, complex drinking water and sewage water systems were organized in Narva–Ivangorod. The delivery of drinking water for Narva inhabitants and a water supply for fire fighters were organized, but the town council decided to start the construction of waterworks. On 28 September 1876 water was pumped to the tanks and the waterworks of Narva started functioning (Nikolayev 1926). The water supply became permanent in the early 1880s (Andreeva *et al.* 2001, p. 46). By the end of the nineteenth century, the population of Narva was 30,000 people, comprising many nationalities, and the town hosted half of Estonia's industrial workers, of whom 5,400 worked at Kreenholm. The town continued to grow, and in 1913 45,000 people lived in the greater Narva area, including Ivangorod. The number of workers at Kreenholm was over 10,000 (Eesti Entsüklopeedia 1994a). The river water thus became commodified through simple technical mediations by people, industries and power stations.

Estonia declared independence in 1918 and signed a peace treaty with Russia in 1920. Accordingly, Ivangorod (renamed as Jaanilinn in Estonian) became a part of Estonia and Narva. Due to the poor relationship between Estonia and Russia, economic activities in the Narva area declined, as did the population, which fell to 22,500. Later, as World War II exploded between Estonia and Russia the frontline was located in and around Narva for some months. The town was devastated: 98.2% of the buildings were destroyed, and the remaining buildings were substantially damaged (Weiss-Wendt 1997).

After the war, Estonia became part of the Soviet Union. The human–non-human network around the water extended spatially, including inhabitants, various local–regional political and economic interest groups, textile and other factories as well as naturalized entities of the environment, etc. Soon after the war, the damaged textile industries were reconstructed. The labor force was brought in from different parts of the Soviet Union, and Russian became the most common mode of communication. Ivangorod and Narva formed a common functional area, and additional connecting bridges were constructed across the river.

The period from the 1950s until the 1990s transformed both the river and local water usage. Until the end of the 1950s, water for the common water supply network was taken from the waterworks and purified according to very basic methods (Andreeva *et al.* 2001). In 1957, a raw water purification plant was opened in Mustajõe, 20 km upriver from the town. The Narva sewage station was established soon after, and a year later, in 1967, it treated half of the waste water from Ivangorod. The other half went without cleaning to the river, creating environmental problems downstream. Narva and Ivangorod were partly functional urban areas, although infrastructure was not fully networked across the border.

The transformation of water into a quasi-commodity brought about the necessity of measuring consumption through infrastructure. Once a month, representatives of Narva and Ivangorod town water organizations jointly measured the processed quantities of drinking water and sewage.[1] The water measurement instrument for Ivangorod was located in Narva. The amount of sewage delivered from Ivangorod to the Narva station was measured with pipe instruments in Ivangorod.

Technical adjustments to water infrastructure required cooperation between the water companies.

Even more significant was the development of specific industries related to water. In the late 1950s, a hydroelectric power station was constructed on the Ivangorod side of the river. The river flow based on natural variation ended, and the riverbed at the local waterfall was dried. To regulate the water quantity, a water reservoir with a dam was built upstream. In 1956, the reservoir reached its size of 191 square km, with a depth up to 15 meters in some parts but with a low depth on average (Eesti Entsüklopeedia 1994b). Regulated water flows were needed for the expanding textile industry and the power station for exploiting energy resources. In the surrounding regions of the Estonian SSR, the mining of oil shale was increased substantially. This material was transformed into energy by two power plants located by the river, the first of which opened in 1959 and the second in 1969. Both needed large quantities of water for this process. However, some industrial plants created their own water circulation system.

The enhancement of water infrastructure dictated a more comprehensive form of management of the water network. The Narva Municipal Office for Water Supply and Sewage was established in 1965 and subsequently modified in 1971 and 1984 under the Soviet Estonian general administration for water supply and sewage. By the end of the Soviet period, Ivangorod had 12,000 and Narva about 89,000 inhabitants living in a significant industrial urban area (Eesti Entsüklopeedia 1994b). The complex water infrastructure spanned from dikes to channels, locks, pumps and pipes.

Water was considered a commodity in Narva–Ivangorod but not significant as a direct resource from which to gain economic profit. There were similarities with the nationalized water industries and the state hydraulic paradigm of water management in Western Europe (Bakker 2005), namely planning for supply-led growth to support economic development, universal provision, and state ownership. However, the Soviet regulatory system was much less concerned with cost recovery and social equity related to the means of the individual and collective consumer. For example, the inhabitants paid a small monthly fee regardless of the actual amount of water used. The water infrastructure was developed to meet the demand for water to feed the five-year plans of the growing textile industry. New pipes were constructed with a view to further expansion of the industrial use of water, in which cost considerations did not play a fundamental role. Here again, one detects some similarities to the Western European context where the use of water infrastructure is promoted to its fullest capacity. However, discussing this question in detail would require a study of its own.

The disintegration of water infrastructure networks

Since Estonia regained its independence in 1991, the Narva River has marked the *de facto* border between Estonia and Russia and, since 2004, also between the European Union and the Russian Federation. However, the *de jure* border is approximately 10 km east from the river as it was in 1920, for, as of 2009, Estonia and Russia have not yet signed a mutually agreed border treaty. Consequently, at present the river divides the urban area into Narva on the western shore and Ivangorod on the eastern shore.

After 1991, the use and ownership of the once common infrastructure across the river became complicated between not only Estonia and the Russian Federation, but also between local, regional and central authorities. For example, the *Sõpruse* ('Friendship') bridge over the river connecting the centers of Narva and Ivangorod had been built for passengers and motor vehicles in 1960. The *de facto* border was established in the middle of the bridge and border stations of the respective states were located on the two sides of the bridge, while the bridge itself became the property of Estonia. In addition, a small pedestrian bridge, *Parussinka*, was owned by the municipality of Ivangorod and a railway bridge was co-owned by the private Estonian Railways and the Russian state authorities.

Functional networks intertwining Narva–Ivangorod soon started to disintegrate, even though the socionatural networks remained. The common public transportation system ceased to exist by 1992 and the common phone system by 1994. In 1996, the Baltic electricity station in Estonia stopped providing heat to one Ivangorod district. Despite these changes, local inhabitants continued to cross the border almost as before to work in textile factories on both sides of the river, to profit from a form of bazaar economy using a simplified border-crossing procedure, or to visit relatives and graveyards on both sides of the river. Drinking water was provided and sewage purified through Narva. Today, a majority (95%) of Narva inhabitants remain speakers of Russian and less than half are Estonian citizens.[2] Some 8,000–12,000 people crossed the border daily before a stronger visa regime was established in 2000 (Berg 2001, pp. 109–15). Until Estonia became a member of the European Union, the number of daily border crossings remained quite the same, being 8,800 people in 2004, but the number of cars had risen to 1,050 (Estonian Border Guard 2005). More recently, queues consisting of hundreds of lorries have been observed waiting to cross the river and the border. Socionatural actor-networks in, around and over the river and water have become more complex. Diverse spatial relations emerged around networked infrastructures between Narva and Ivangorod. The official border was manifested and crossed through these links. Nevertheless, the administrative borders did not overlap with the transboundary circulations and the ownership of infrastructures.

The period following 1991 has been marked by the disintegration of various water-related material and human networks between Narva and Ivangorod. The water supply and sewage infrastructures were shared and used equally by Narva and Ivangorod despite the political separation between Estonia and Russia. Urban networked infrastructures and related circuits were bundled together (Graham & Marvin 2001), and for a while they resisted wider political changes. Urbanized water in the form of drinking and sewage circulations is the very basis for urban living. So it was necessary to find temporary (geo)political solutions to strategic changes in infrastructure links.

Reorganization of Estonia's water network began soon after independence with the foundation of Eesti Vesi, a state-owned enterprise to which Narva Vesi also belonged. In 1993, Narva Vesi became a municipal enterprise, and in 1995 a public limited company owned by the municipality (Narva Vesi 2007).[3] Continued public ownership of the water delivery system later made it possible to draw down European Union funding for the water delivery system.

The nearby town of Narva-Jõesuu also became associated with the ownership of this non-profit company. The distribution of dividends was not the object, and only limited municipal investment was expected. The investment plan was, from the beginning, based on receiving financial aid from the European Union, expected to be up to three-quarters of the total infrastructure investment costs. Narva Vesi's finance plans framed the local water prices in Narva and Ivangorod (Rist 2006). The Narva–Ivangorod case is different to many of the privatization projects of the 1990s in Africa, Asia and Latin America. There the main aim was to achieve greater efficiency and expansion in the water and sanitation sector. However, despite its prominence in the water sector, the scale and benefits of private-sector participation remained limited. In the early 2000s only around 5% of the world's population had their water provided by the private sector; in Western Europe, however, this figure reaches 25% (Budds & McGranahan 2003, pp. 87–88; Davis 2005).

In the 1990s, water pipes physically connected households and factories on both sides of the river. In the authors' interviews in Narva during 2005, this water infrastructure network was described as 'one organism'. The point for distribution of drinking water to Ivangorod was on the Estonian side, where the amount of water provided to the Russian side was measured. The delivery of drinking water to Ivangorod could be regulated in Narva. The sewage of Narva and Ivangorod first went through the Narva purification station and was thereafter directed down the Narva River (Sizova & Sizov 2000, p. 19). Water pipes with a diameter of 1.2 meters ran between Narva and Ivangorod on the riverbed close to the *Parussinka* pedestrian bridge. The technical condition of the pipes was monitored every 10 years, and in 1997 the pipes were considered to be in good shape and the remaining exploitation time was estimated to be up to 20 years. These pipes belonged, and still belong, to the municipality of Ivangorod, which manages the pipes.[4]

Narva Vesi became responsible for water supply to both Narva and Ivangorod. The socio-technical transformation of water infrastructures and water networks became a new issue and particular human and nonhuman networks assembled and disassembled around drinking and wastewater. The Narva Vesi management agents raised their power position in the network. The municipality of Ivangorod distributed water within the town and paid monthly for the amount consumed.

The relevant authorities of the two towns held a few annual meetings in the early 1990s. Otherwise there was little cooperation and, even then only in practical matters. In a 2005 interview the Narva Vesi Finance Director for 1997–2004, Mr Rist, noted that:

> [The municipal enterprises for the water provision in Narva and Ivangorod] did not really communicate. When they [in Ivangorod] informed us that the [water] pressure was low, the pressure at the pumping house was raised a bit. Once a month the measures [of water consumption] were taken, the bill presented and then – nothing more. We [in Narva] sold [water] wholesale to the municipal enterprise, which distributed it further to the people. Once a month we sent the bill and took it for granted that the money would arrive. However, it did not and the problem began from there.[5]

The network relations between two crucial actors in the distribution of water became formal. The market-oriented institution in Narva started to gain prevalence.

However, Narva Vesi was not able to completely commodify water, nor reassemble the socionatural network on market-driven principles.

Towards the second half of the 1990s, the water issue started to change substantially. Narva Vesi continued to distribute water to Narva and Ivangorod. However, certain cost-related aspects were on the agenda on the Estonian side. New metering systems of water consumption were introduced during 1994–1995 in Estonia in general and especially in Narva. The inhabitants were motivated to install household water consumption meters, as they would thus pay less for water consumption. The daily water consumption norms of the Soviet Union per person (300–400 liters) were about three times higher than the actual consumption, which is why the water norm made household water bills high. In 1996–1997, Narva Vesi implemented a European Union-funded project to install water consumption meters in Narva households.[6] The less wealthy Narva inhabitants got their meters through loans. The household water meters resulted in a decrease in actual water consumption and, subsequently, a decrease in the income of Narva Vesi, and thus created a need to raise the price of water. Water consumption was first paid for by the inhabitants to the housing cooperative in which they lived and, subsequently, by the cooperative to Narva Vesi. However, the organization of households into housing cooperatives was very slow in Narva. In addition, large amounts of water were lost due to leaking water pipes and the deliberate theft of water. Narva Vesi was unable to use the meters of the housing co-operatives and thus had to deal directly with households.

The rationalization and marketization of water services became further issues, affecting the development of water networks in Narva–Ivangorod. In Ivangorod, water was paid for according to a fixed monthly fee subsidized by the Russian state authorities up to 50%. However, Narva Vesi raised the water price approximately nine-fold during the 1990s.[7] The former finance director of Narva Vesi explained that the water price was set in accordance with the company's financial plan.[8] It was first discussed and approved by the Narva Vesi board, which consisted of political representatives of the Narva city council. The proposal was then discussed, frequently amended, and finally approved by the Narva city council. Those members of the Narva Vesi board who were on the council exerted pressure to keep the water price low because Narva's inhabitants did not want to pay more. A low price was thus also in the interest of the leading political party in Narva. However, such a pricing policy reduced the budget of Narva Vesi and lessened their investment in water infrastructure. The water price for Ivangorod was lower because the wholesale price did not have to include delivery infrastructure costs. These networks between actors show that economic and political interests were closely intertwined within the decision-making process concerning the price of water in Narva and Ivangorod.

The most rapid change in price occurred during 1997–1998, after which the cost was triple the standard price in the Leningrad *Oblast'*. Nevertheless, the Russian authorities continued to subsidize water use in Ivangorod as before. Soon the municipality of Ivangorod became indebted to Narva Vesi because it and its inhabitants were not able to pay for '*the golden water of Narva*' (Tsvetskov 2005).[9] Due to the increasing debts of Ivangorod, Narva Vesi introduced periodic cuts in 1997–1999, which involved not providing drinking water to Ivangorod or providing it for just a few hours per day and with low pressure. These periodic cuts aimed to push the municipality of Ivangorod, or other Russian authorities, to pay their increasing debts to Narva Vesi. However, Narva city council did not go so far as to cut off the water supply to Ivangorod altogether.

During 1997 and 1999, Narva Vesi informed the authorities in Ivangorod and the Estonian media in advance about the closure of the water supply. The Ivangorod municipality then informed the inhabitants, so that people were able to collect water in advance. However, regardless of whether they received water or not, Ivangorod inhabitants had to pay the monthly fixed water fee to the local authorities. This clearly illustrates the imbalance of power positions within the water network. The access of local people to drinking water in Ivangorod was not bound to personal economic wealth but rather to the location of residents' apartments. The people living in higher floors of apartment blocks in Ivangorod remained without water. In Ivangorod, various temporary water networks and alternative ways to get drinking water bypassing the water commercialization by Narva Vesi emerged. Looking back, in 2005 many inhabitants of Ivangorod remembered the water supply disruptions lasting from a few days to several months. During these periods, they got water from various sources: street artesian wells, summer cottages, organized water tanks near their home, factory buildings, etc. These alternative ways to get drinking water established particular short-term practices.

The water issue also reached prominence in international politics and it was actively discussed in the national and international media. This brought new and spatially more distant actors to the network. Some Russian authorities implied that the Estonians had disrupted the water supply for political reasons and threatened retaliation. For example, in 1998, the Leningrad *Oblast'* authorities threatened to stop delivering gas to Estonia, and political activists were going to disconnect the railway between Russia and Estonia, but these counter-measures were never realized. In addition, one political activist from Estonia invited Ivangorod to join Estonia but this announcement was made in order to attract more media attention to the water issue.

In addition, the government of Estonia became actively involved. In 1997, it paid without broader consultation a part (about €585,000) of Ivangorod's debt to Narva Vesi as part of humanitarian aid from Estonia to neighboring countries. This marked the first ever instance of Estonia directing humanitarian aid to Russia. This gesture, made during the Russian economic crisis, reached the public through a caricature in the biggest Estonian daily newspaper *Postimees* (15 March 1997) which stated that: 'Giving irredeemable financial support to Russia is now in fashion'. The Republic of Estonia gave humanitarian aid to Russia through the Estonian local municipal water company. The aid was directed towards Ivangorod inhabitants to provide them with access to drinking water. Simultaneously the Estonian municipal water company avoided bankruptcy and the loss of local jobs. In the process of water commodification one municipal company gained visible power and water as the resource became intertwined with wider international (geo)politics.

By the end of 1998, the debts of Ivangorod to Narva Vesi had risen to over €1 million and the Leningrad *Oblast'* had not been able to reduce it. The Estonian government paid an additional €70,000 of the Ivangorod water service-related debts to prevent the dumping of sewerage from Ivangorod into the river,[10] thus bringing into focus the public necessities regarding water, sanitation and drainage, necessities that are often underestimated when a private company rationalizes the price of water services (see also Budds & McGranahan 2003, p. 96; Davis 2005). Since avoiding sewage dumping by Ivangorod town into the Narva River was obviously in the public interest both of Narva's inhabitants and of the wider collective. By 1998, Narva Vesi no longer took waste water from Ivangorod. Half of the Ivangorod waste water was cleaned and the remainder was directed without cleaning to the Narva River below the town. This obviously created annoyance and environmental concerns.

There were also local–regional attempts to solve the water issue. Several Russian politicians and companies promised to sort out the complicated situation with Narva Vesi. Some private Russian companies proposed to mediate payment of debts with other items, such as oil shale, coal and wood, as had been common in the Soviet Union. Only a few real efforts were made and all of them failed. In fact, such negotiation about water payments was not a new issue in Narva–Ivangorod. Already during the Soviet period Ivangorod had sometimes become indebted by not paying for water in time. At that time this was solved locally by, for example, paying the debt with other commodities, i.e. giving certain products to the water management office in Narva or by finding temporary help from a ministry dealing with such issues.[11] Barter (exchange of commodities, no money involved) was quite common between state-owned enterprises during the Soviet era. However, after 1991 Narva and Ivangorod were in two different countries, which made such local, and sometimes non-transparent, agreements difficult. Narva Vesi required the (measured) water debts back in the form of money. According to Appadurai (1986, p. 13), 'in trying to understand what is distinctive about commodity exchange, it does not make sense to distinguish it sharply either from barter on the one hand, or from the exchange of gifts on the other'. It is important to see the calculative dimension in all these forms of exchange. Water becomes an exchangeable commodity for something else. Therefore Appadurai's (1986) reflection is relevant to the current case study and extends commodification to the pre-capitalist period.

The largest Estonian daily *Postimees* published about 20 short articles about the water conflict during 1996–1998. The main issue was the (non)payment for water by the municipality of Ivangorod. The Mayor of Narva, the Consul of the Leningrad *Oblast'*, the Manager of Narva Vesi, and the Prime Minister of Estonia were seen as the key actors. Some articles conflated the water issue with the broader issue of relations between the Republic of Estonia and the Russian Federation. On 12 November 1998, a *Postimees* editorial noted ironically that:

> 'Estonia is not as yet rich enough to clear the debts of the big neighbor, especially not when the latter points with a finger to our problems with human rights [regarding the position of ethnic Russians in Estonia]. There is no obligation to drink the water of a state in which your compatriots are being discriminated against'. (Translation from Estonian by the authors)

This statement illustrates the symbolic position given to the everyday consumer in Ivangorod: water and related debts are associated with wider geopolitical questions, and material and semiotic relations became mixed within the water network. The usual commodity exchange of drinking water and sewage circuits were negotiated through gift exchange and barter agreement. The earlier humanitarian aid from

Estonia to Ivangorod was a gift. Although an exchange of gifts from the Russian side was not directly expected, a feeling of possible reciprocity and better political behavior was created.

However, the decisive turning point came in 1999 when the debts of Ivangorod to Narva Vesi had risen to over €1.2 million. The director of Narva Vesi (Ers 2000, p. 23) wrote that:

> This year [1999], Narva Vesi has decided to terminate its business contacts with its very substantial client because Jaanilinn [the name of Ivangorod in Estonian] has not managed to pay its debt for water since 1992[?]. Today, Jaanilinn is not short of drinking water thanks to the usage of artesian wells built in 1995. However, [that Ivangorod is] discharging practically unclean wastewater into the Narva River is not the best solution. (Translation from Estonian by the authors)

This indicates how environmental concerns were articulated alongside money-related issues in the disintegration process of the socionatural water network. The metabolisms of drinking water were not directly connected to the unclean sewage leaks to the Narva River, although the artesian wells in Ivangorod were not very deep at the beginning of the period of water cuts and sewage leaks. Unclean sewage flows also became the center of attention at the international level because unclean sewage water ended up in the Baltic Sea. Nevertheless, the directing of waste water into the shared river indicates that the public benefits of water and sanitation were underestimated by Narva Vesi in the rush to increase service charges and demand payment of growing debts by Ivangorod/Russia.

The issue of solving the water debts continued for some years after the water delivery was terminated. In 2001, Narva Vesi sued the Ivangorod municipality in a Russian court, which ruled in favor of Narva Vesi. The Leningrad *Oblast'* gradually paid the water debts to Narva Vesi in 2001–2004. In the authors' interviews held in Narva in 2005, the financial debts of Ivangorod and the different economic systems between Estonia and Russia were mentioned as the main reasons for the curtailment of drinking water delivery and wastewater collection. In Ivangorod, the local authorities referred also to broader political reasons.

The commodification of water during the 1990s was mostly around pricing issues. When the public sector provides scarce consumables free (or at subsidized prices), people have an incentive to overuse them (Budds & McGranahan 2003, p. 95). The drinking water in Narva and Ivangorod was not very scarce because it was possible to temporarily acquire water from additional sources and use the river for sanitation drainage. Nevertheless, the increasing price of water services and installed water meters significantly decreased the amount of water consumed in Narva. Currently the water consumption is about one-third of what the water infrastructure was built for. According to European Union regulations, full-cost pricing must be accompanied by measures that secure water use for everyone, but without resulting in over-consumption at the aggregate level (Bithas 2008, p. 1299).

The commodification of water and the increase in water service prices have been a general tendency in Estonia, as in many Western countries. The Prime Minister of Estonia, Andrus Ansip, expected in fall 2007 that the water price would rise substantially in towns in Estonia (Kalamees 2007). Despite a rapid and substantial

(46%) water price rise in 2008 (Narva Vesi 2008), water costs in Narva are still below the Estonian average. The water economy in Narva is still mostly supply-led without competition, and local politics intervenes in price regulation in ways that are difficult to forecast. At the time of writing, Narva Vesi does not have a clear and transparent price regulation system, and this fact has been criticized by the Estonian ombudsman (Taras 2007). The rise in water price as proposed by Narva Vesi is still met with public opposition in Narva. According to critics, the rise is not justified, as Narva Vesi was not originally established to make profit. In addition, the rise would cause problems for local industries and, hence, to local employment.[12]

In Ivangorod, meanwhile, drinking water continues to be provided by the municipal enterprise of Ivangorod. The water is paid for according to a fixed monthly fee subsidized by the Russian State. The fee is not related to the actual consumption by the inhabitants, nor is the consumption measured in the households. Similarly to the heating system in Russia, the local water systems are 'technologically collective – it is impossible to regulate individual use or shut off individual households' (Collier 2004, p. 52). In Narva, there is a modern water network in which consumption is measured and the price paid is according to the market situation. However, the price of water has risen so much that the reduction in the use of water in the past ten years does not mean a reduction in the actual costs for the inhabitants.

Narva Vesi has worked with a consultancy firm to improve the water management system. Its report suggests steps (e.g. efficiency criteria, outsourcing) for the commercialization of water supply in Narva and Narva-Jõesuu (Business grain 2002). However, as Bakker (2005) has observed with regard to England and Wales, water companies remain vertically integrated monopolies and full commodification and commercialization of water has been incomplete. The water economy in Narva and Narva-Jõesuu is still mostly supply-led without competition. The public ownership of Narva Vesi has forced the company to keep the water price relatively low. Currently, Narva Vesi does not have a clear price regulation system and does not take into account the amortization costs in the investment plan. The full privatization of Narva Vesi has been postponed by ten years due to EU requirements regarding resource use. As Budds & McGranahan (2003, pp. 111–12) argue on the basis of water studies in Africa, Asia and Latin America, private-sector involvement does not result in more equal water distribution, Nor does it eliminate the politicization of water and sanitation provision – indeed, it can even heighten them and bring in new forms of corruption. Recent trends indicate that the rate of privatization has been slowing since the late 1990s, due to a combination of the underestimation of risks, the overestimation of profits and problems with contracts in some cases.

Building up parallel socionatural water networks

The physical water infrastructure network was improved in various ways during the 1990s, but often separately by Narva and Ivangorod. Such improvement aimed to stabilize water supply and sewage treatment into new and separated water networks. Narva Vesi headed towards water supply commercialization, while the Ivangorod municipality continued with state-subsided water. Faced with the emerging political border between Narva and Ivangorod, the municipality of Ivangorod took the initiative to make changes to their water networks. The construction of new local water infrastructure started in the early 1990s, and was half-finished by the time of the final water closure in January 1999. A parallel socionatural network was in construction without including the Estonian partners. New water bores were drilled to allow the supply of low-quality drinking water from local artesian wells for some Ivangorod inhabitants. Gradually the municipality of Ivangorod improved its own water infrastructure and built a sewage station, which was opened in 2001. The locations of the Ivangorod water and sewage stations are formally kept a secret, but their locations are known to specialists in Narva, thus making Estonians passive participants in the new water network.

The water infrastructure of Narva also changed during the late 1990s. More than three-quarters of the old water infrastructure in Narva, built at the end of 1960s, had been deteriorating for various reasons, including limited maintenance and diminished use. At times, there were 15–17 simultaneous leaks undergoing repair and a lot of water was lost (Ers 2000). In 2002, there were 132 pipe leaks in the drinking water network and 16 leaks in the sewage infrastructure (Tereblina & Semyanov 2002). Narva Vesi also closed down most of the 21 artesian wells, of which three were integrated into the municipal water network. Four artesian wells belonged to industrial plants and remained partly in use. The reason for the closure of the artesian wells was that, according to a survey, there was a lack of water. The cities of St Petersburg, Kingisepp and Ivangorod used water from the same geological layer.[13] Nevertheless, the artesian wells can still be re-opened if water is urgently needed. From the early 2000s almost all houses in Narva have been equipped with a water measurement system and the Narva inhabitants pay for the water they consume (Ers 2000).

The socionatural network has become increasingly technologized, and certain nonhuman technology actors have replaced human actors. Regarding the sewage infrastructure in Narva, in 1997–1998 a new purification plant replaced the previous one in Mustajõe and a new pipe from the Mustajõe River to the Narva town was put into operation. Since 2002, the Mustajõe water station has been automatic and without staff. The water quality is measured weekly and technical control is conducted on a monthly basis. Access to the station requires a special permit because it is located in the border zone.[14]

Today the drinking and waste water infrastructure in Narva and Ivangorod run parallel with other networks formed along the flowing water. Recently some international actors have infiltrated the water networks in and between Narva and Ivangorod. Danish and Swedish specialists conducted feasibility studies to reconstruct the water network and sewage cleaning plant in Ivangorod (Ers 2000). In addition, there was a proposal to integrate the towns of north-east Estonia with the water system of Narva. Municipalities looked for suitable scales for water service management. The preparatory project was mostly financed by loans from the European Bank for Reconstruction and Development, with 10% coming from the Narva municipality. Finally, only Narva-Jõesuu and Narva took part in the project. The European Union subsidized the Narva sewage networks with over €9.2 million. At the end of 2005, Narva Vesi opened a new sewage station, which fully replaced the one from 1960. Besides lessening chemical water pollution, the modern sewage station technology also decreased air pollution. The funding came from the European Union (56%), the Swedish Agency for Investments and Development (19%), the Estonian Environmental Investment Center (15%), and Narva Vesi (10%). The station was constructed by the Danish–Swedish–Estonian consortium of Skanska EMV, Krüger A/S and Purac AB (Narva Water 2006). The development plan for Narva indicates that an investment of about €27 million is required to reconstruct the common drinking and waste water infrastructure systems of Narva, Sillamäe and the nearby villages of the Vaivara municipality from 2009 to 2015.[15] This shows how infrastructure is reorganized following the economics of scale to find regionally suitable solutions for networks.

The socionatural water networks of Narva–Ivangorod have been disintegrating for more than a decade. Today, Narva and Ivangorod are still connected by the material water infrastructure and human–non-human water networks exist around it. However, the common water pipes between Narva and Ivangorod are not in use and are partially out of order. The organizations in Narva and Ivangorod responsible for drinking and waste water are neither in conflict nor in cooperation. Cooperation is possible if, for example, the water infrastructure needs to be urgently repaired. However, both organizations argue that there should be no return to the previous common infrastructure system. The water organizations in Narva and Ivangorod both confidently claim that they provide better water with a lower price for their customers.[16]

Emerging issues: health and environment

Water consumption by inhabitants has diminished radically in both Narva and Ivangorod since the Soviet Union because the population has fallen, industries have ceased their activities and rising costs have lessened the use of water. In 2006, Narva Vesi provided the town with 4.43 million cubic meters of water along its 161-km long pipe network and treated 6.89 million cubic meters of sewage water along its 201-km long sewer system. The difference between these two numbers is due to several industries in Narva having their own water bores. The amounts of water delivered and waste treated have decreased since 2004, although the capacities of water and sewage treatment stations have increased. The largest client groups of the Narva Vesi water delivery are the local inhabitants (2.6 million cubic meters), Kreenholm textile factory (0.2 million cubic meters) and nearby Narva-Jõesuu town (0.1 million cubic meters). Kreenholm has its own water bores, but its large amounts of waste water (1.4 million cubic meters) are treated by Narva Vesi.[17] However, Kreenholm is struggling to avoid bankruptcy and its economic activities are being reduced.

While households and local industries are significant users of water, the Baltic electricity station located by the river is by far the largest consumer, using water to cool industrial processes. Its industrial water originates from the water reservoir located upstream of the Narva River. The water goes through several cleaning phases in the electricity station, and after use mostly goes back to the reservoir. When taking into account this industrial use of water, almost 30% of all consumed water in Estonia is used in Narva. However, 96% of the total water consumption in Narva was used in

1998 directly by the Baltic electricity station.[18] This underlines the importance of the upstream water reservoir built in 1956 to regulate the day–night water flow for hydroelectric power stations. Furthermore, the water reservoir became a water resource and cooling water reservoir for two large electricity power stations. In 2007, these power stations were running at below capacity.[19] In recent years, there has often been discussion about the threat that the Russians would empty the reservoir. This would have a dramatic impact on Estonian energy supply, in which the power stations in Narva play a fundamental role. However, to discuss this at length would require an additional study.

New issues have arisen as regards water networks in Narva and Ivangorod, namely those related to health and the environment. In 2005, some Ivangorod inhabitants reported that the drinking water was not of good quality, possibly owing to the old water infrastructure connecting the houses. The varying quality of drinking water is also a problem in Narva, especially during the spring floods. In addition, many Narva inhabitants reportedly consume bottled water for everyday use. In fact, a health-related debate about the water quality in Narva exploded in late 2007. The national health authorities forbade the selling of tap water by Narva Vesi. The reason was that according to the European Union regulations, the drinking water contained an undue quantity of a particular chemical substance.

By early 2009 a solution to the possible health risks caused by the tap water in Narva had not been found. However, the managers of Narva Vesi claim that they do not need permission from the national health authorities to deliver water. Nevertheless, European Union water quality standards have encouraged Narva Vesi to develop ties among actors who must take account of each other's interests and influences. These processes are common in Eastern Europe (see Bresssers & O'Toole 1994). The situation of technical water quality in Ivangorod is unknown. The European Union regulations do not apply there, so the Estonian authorities do not have the legitimacy and the Russian national authorities the necessity to pinpoint the possible risks of drinking water. In the privatization of water services, a major concern is that they are so central to public and environmental health that responsibility for their provision cannot be entrusted to for-profit organizations (Davis 2005, p. 146).

The water economy in Narva has several negative environmental impacts, but these have not been discussed much so far. One issue is waste water leaks into the river and the water reservoir. In addition to waste water from households, other important sources of pollution are thermal and alkaline waste from the electricity power station as well as indirect polluters, such as the Narva rubbish-collection plant.[20] There is no clear evidence as to whether private-sector participation in the water supply and networks is better or worse as regards the environment. It is particularly relevant whether the utility is compensated or charged for the value of externalities it produces or avoids (Davis 2005, pp. 160–4). Environmental questions have received attention in Narva through water and sewage infrastructure projects. Obviously, the political and economic context strongly influences how much the provider Narva Vesi takes the environment into consideration. There have also been calls to restore the rapids in the River Narva on environmental grounds. This would have an impact on the river as an energy source for the power plants.

In Narva–Ivangorod there has not yet been a shift to market environmentalism in the management of the water supply. The position of the environment (cost, nature) has been weak, price competition absent and resource costs have been insignificant on the water agenda. In general, the private water companies in pursuit of profit maximization are likely to ignore the relevant external costs such as those regarding environmental protection and user/scarcity costs (Bithas 2008, p. 1303). However, as Bakker (2005) has observed, in the OECD countries market environmentalism is on the rise as regards water supply and networks.

Conclusions

Considering water networks as a socionature means that one does not separate human activities (society) and environment (nature). This perspective gives new insights into the commodification of water resources, including hybrid human and non-human entities. Without water, there is no Narva–Ivangorod. Water also marks the current division between Narva and Ivangorod in the wider socio-political context. The water network in Narva–Ivangorod emerged, changed, remained united, and split into two directions that are still partly bound together. The transboundary context of water networks adds complexity and encourages analysis of the existing transboundary water links and circulations. Three major conclusions emerge from the case studied.

First, the establishment and renovation of transboundary water infrastructure demonstrates that water networks are socionatural, involving specific technologies and skills. New technologies were established to extend water networks to cover and bind together Narva and Ivangorod. The natural water resource was treated and transported from a distance, and specialists were trained to manage the water infrastructure system. Social networking was needed to secure and regulate transboundary drinking water and sewage circulation. The distribution grids and topological connections linked systems and practices of production and consumption. The water network connects and disconnects places and configures social relations (see Graham & Marvin 2001).

Second, a study of water as a commodity (Appadurai 1986, p. 6) demonstrates how, for instance, tap water is a socialized and urbanized 'thing' that embodies certain values. An exchange sets the parameters of utility and scarcity and is also a source of value. The creation of value for natural resources is a socionatural process intertwining human and non-human actors (see Whatmore 2002). The transformation of transboundary water circulations in Narva–Ivangorod includes complex actor networks with regard to the commodity value of drinking water and sewage. Through money and capturing the agencies and powers of human and non-human actors, some actors have more capacity to direct socionatural relations than others (see Castree 2002). Commodification is important in the transformation of transboundary water flows and infrastructure. The transformation is persistently diverse and the diversity arises out of such multi-scalar relations (Castree 2003). In the case under study, the shifts first to the Soviet Union command economy and later to the European Union market economy and its water management policies were crucial changes. Water became a measured and monetarized commodity within transboundary circulations.

The transformation of water networks in Narva–Ivangorod is similar to the neoliberal commodification of water resources in Eastern Europe and developing countries (see Bakker 2005; Budds & McGranahan 2003; Davis 2005). However, in Narva–Ivangorod water services have not been privatized and equal water distribution has not been so intensively on the agenda compared, for example, with Latin American cities. The politicization of transboundary drinking water and water sanitation services has been clearly evident in Narva–Ivangorod. More recently, new regulations and funding from the European Union have provided pressures and possibilities for local companies to improve the quality of water services. Initiatives to reconstruct water infrastructure were planned towards the scale economics of drinking water and waste water treatment services.

Third, the politics of transforming water networks illustrates how water infrastructure and related (commodity) circuits used to be hidden. They became visible only when links broke down and things did not reach agreed destinations (Law 2000). In the process of integration, disintegration and reintegration of water networks, certain actors and things became more visible and powerful while others were hidden or pushed backwards. The study supports the argument of Bruno Latour (2005), who maintains that it is not society that holds things together, but rather society itself that is held together by things. The positions of agencies were redefined, renegotiated and rebordered through material-semiotic relations to water in particular situations. This became evident in the ruptures of water infrastructure networks between Narva and Ivangorod – when water circulations did not reach the usual 'addresses' and different (commodity) values of water were under negotiation. The drinking water cuts and sewage leaks into the river were socionatural manifestations of transforming transboundary relations. The water cuts and sewage dumping also raised discussions about the water quality and environment renegotiating boundaries of society and nature. Rising water price, planned out-sourcing, efficiency indicators, and a shift towards market environmentalism are general trends of commodification and commercialization of local water services, and these are increasingly evident in Narva–Ivangorod. This relates to a more general neoliberal commodification of nature but with glocal peculiarities.

Notes

1 Authors' interview with S. Tsvetkov, head of client service department, Narva Water, and S. Malihhin, infrastructure expert, Narva Water, 19 January 2005, Narva.

2 Narva Municipality Board, Narva Development Plan 2008–2010, available at: http://www.narva.ee/index.php?vasakul/juhtimine/areng/, accessed 23 October 2007.

3 Narva Water, 'History', available at: http://www.narvavesi.ee/96est.html, accessed 12 September 2007.

4 See note 1.

5 Authors' interview with T. Rist, Narva Water finance director from 1997 to 2004, and Narva Electricity's development manager from 2005 to 2006, 18 January 2005, Narva.

6 See note 5.
7 See note 1.
8 See note 5.
9 See note 1.
10 Estonian Ministry of Foreign Affairs, Estonian humanitarian and development aid in 1998, available at: http://www.vm.ee/est/kat_425/3220.html, accessed 5 January 2006.
11 See note 1.
12 See note 5.
13 Narva Department of the Architecture and City Planning, 'Narva linna üldplaneering 2000–2012. Ülevaade olemasolevast olukorrast' ['The master plan of Narva 2000–2012. Overview of the current situation'], available at: http://www.narvaplan.ee/uldplan/I%20koide.doc, accessed 15 March 2005. See also authors' interview with S. Tsvetkov, head of client service department, Narva Water, and S. Malihhin, infrastructure expert, Narva Water, 19 January 2005, Narva.
14 See note 1.
15 See note 2.
16 Authors' interview with A. Shanin, Ivangorod municipality, first deputy of the municipality formation on issues of local authority and municipal services, 21 January 2005, Ivangorod. See also authors' interview with S. Tsvetkov, head of client service department, Narva Water, and S. Malihhin, infrastructure expert, Narva Water, 19 January 2005, Narva.
17 Narva municipality, 'Narva arvudes, 2006' ['Narva in figures, 2006'], available at: http://www.narva.ee/index.php?vasakul/statistika+Narvast+2006/, accessed 1 November 2007.
18 Narva Department of the Architecture and City Planning, 'Narva linna üldplaneering 2000–2012. Ülevaade olemasolevast olukorrast' ['The master plan of Narva 2000–2012. Overview of the current situation'], available at: http://www.narvaplan.ee/uldplan/I%20koide.doc, accessed 15 March 2005.
19 See note 2.
20 See note 18.

References

Andreeva, C., Bulatnikov, V., Vester, A., Kabadeitseva, N., Klimina, T., Sinyakova, G., Smirnova, G., Smolokurov, A., Solodova, I. & Chudinova, M. (2001) *Narva. Kulturno-Istorichesky Spravochnik* [*Narva—Cultural-Historical Guide*] (Narva, Narva museum).

Appadurai, A. (1986) 'Introduction: Commodities and the Politics of Value', in Appadurai, A. (ed.) (1986) *The Social Life of Things. Commodities in Cultural Perspective* (Cambridge, Cambridge University Press).

Bakker, K. (2001) 'Paying for Water: Water Pricing and Equity in England and Wales', *Transactions of the Institute of British Geographers NS*, 26, 2.

Bakker, K. (2005) 'Neoliberalizing Nature? Market Environmentalism in Water Supply in England and Wales', *Annals of the Association of American Geographers*, 95, 4.

Berg, E. (2001) 'Conclusions: "Border-Crossing" in Manifest Perceptions and Actual Needs', in Berg, E. (ed.) (2001) *Negotiating Borders of Multiple Meanings. Research Reports* (Tartu, Peipsi CTC).

Bithas, K. (2008) 'The European Policy of Water Use at the Urban Level in the Context of the Water Framework Directive. Effectiveness, Appropriateness, and Efficiency', *European Planning Studies*, 16, 9.

Braun, B. (2004) 'Nature and Culture: On the Career of a False Problem', in Duncan, J. S., Johnson, N. C. & Schein, R. H. (eds) (2004) *A Companion to Cultural Geography* (Blackwell, Oxford).

Bressers, H. & O'Toole Jr, L. J. (1994) 'Networks and Water Policy: Conclusions and Implications for Research', *Environmental Politics*, 3–4, pp. 197–217.

Budds, J. & McGranahan, G. (2003) 'Are the Debates on Water Privatization Missing the Point? Experiences from Africa, Asia and Latin America', *Environment and Urbanization*, 15, 2.

Business grain (2002) *AS Narva Vesi juhtimissüsteemi analüüsi lõpparuanne [Final report of the Narva Water Ltd management system analysis]* (Tallinn, Business grain).

Castree, N. (2002) 'False Antitheses? Marxism, Nature and Actor-Networks', *Antipode*, 34, 1.

Castree, N. (2003) 'Commodifying What Nature?', *Progress in Human Geography*, 27, 2

Castree, N. & Macmillan, T. (2001) 'Dissolving Dualisms: Actor-Networks and the Reimagination of Nature', in Castree, N. & Braun, B. (eds) (2001) *Social Nature. Theory, Practice and Politics* (Oxford, Blackwell).

Collier, S. J. (2004) 'Pipes', in Harrison, S., Pile, S. & Thrift, N. (eds) (2004) *Patterned Ground: Entanglements of Nature and Culture* (London, Reaktion).

Davidson, D. & Frickel, S. (2004) 'Understanding Environmental Governance: A Critical Review', *Organization and Environment*, 17, 4.

Davis, J. (2005) 'Private-Sector Participation in the Water and Sanitation Sector', *Annual Review of Environment and Resources*, 30.

Eden, S., Tunstall, S. & Tapsell, S. (2000) 'Translating Nature: River Restoration as Nature-Culture', *Environment and Planning D: Society and Space*, 18, 2.

Eesti Entsüklopeedia (1994a) 'Kreenholm', in *Eesti Entsüklopeedia* (1994) (Tallinn, Eesti Entsüklopeediakirjastus).

Eesti Entsüklopeedia (1994b) 'Narva', in *Eesti Entsüklopeedia* (1994) (Tallinn, Eesti Entsüklopeediakirjastus).

Eesti Entsüklopeedia (2003) 'Narva', in *Eesti Entsüklopeedia 12* (2003) (Tallinn, Eesti Entsüklopeediakirjastus).

Ers, A. (2000) 'Water Supply and Sewage Purification of the Narva city', in Jaani, A. (ed.) (2000) *The Narva River and Reservoir* (Narva, Peipsi CTC).

Estonian Border Guard (2005) *Border-crossing statistics* (Tallinn, Estonian Border Guard).

Finger, M., Tamiotti, L. & Allouche, J. (2006) *The Multi-Governance of Water* (New York, University of New York Press).

Graham, S. & Marvin, S. (2001) *Splintering Urbanism. Networked Infrastructures, Technological Mobilities and the Urban Condition* (London, Routledge).

Kaïka, M. (2005) *City of Flows. Water, Modernity, and the City* (London, Routledge).

Kalamees, K. (2007) 'Peaminister Ansip peab vee henna tõusu paljudes omavalitsustes paratamatuks' ['Prime minister Ansip holds that in many local authorities the water price rise is unstoppable'], *Postimees*, 29 October.

Kallis, G. & de Groot, H. (2003) 'Shifting Perspectives on Urban Water Policy in Europe', *European Planning Studies*, 11, 3.

Kortelainen, J. (1999) 'The River as an Actor-Network: The Finnish Forest Industry Utilization of Lake and River Systems', *Geoforum*, 30, 3.

Latour, B. (2005) *Reassembling the Social: An Introduction to Actor-Network Theory* (Oxford, Oxford University Press).

Law, J. (2000) 'Objects, spaces and others' (Lancaster, Department of Sociology, Lancaster University), available at: http://www.comp.lancs.ac.uk/sociology/papers/law-objects-spaces-others.pdf, accessed 15 March 2005.

Nikolayev, P. (1926) 'The beginning of the water management in Narva', available at: www.narvavesi.ee, accessed 9 June 2008.

Rist, T. (2006) 'Kas 2 + 2 on ikka 4?' ['Is 2 + 2 really 4?'], *Kreenholmi Prospekt*, available at: http://www.prospekt.ee/narvavesi.doc, accessed 16 August 2006.

Sizova, T. & Sizov, E. (2000) 'Water Management of the Narva Region', in Jaani, A. (ed.) (2000) *The Narva River and Reservoir* (Narva, Peipsi CTC).

Smith, N. (1984) *Uneven Development. Nature, Capital and the Production of Space* (Oxford, Blackwell).

Swyngedouw, E. (1999) *Modernity and Hybridity—The Production of Nature: Water and Modernisation in Spain*, available at: http://www.geo.ubs.ca/papers/Swyngedouw_E.html, accessed 15 May 2005.

Swyngedouw, E. (2004) 'Water. Circulating Waters, Circulating Moneys, Contested Natures', in Harrison, S., Pile, S. & Thrift, N. (eds) (2004) *Patterned Ground: Entanglements of Nature and Culture* (London, Reaktion).

Swyngedouw, E. & Kaïka, M. (2000) 'The Environment of the City . . . or the Urbanisation of Nature', in Bridge, G. & Watson, S. (eds) (2000) *A Companion to the City* (Oxford, Blackwell).

Taras, M. (2007) 'Jõks: Narvas tõsteti põhjendamatult vee hinda' ['Jõks: Water price was risen without ground in Narva'], *Eesti Päevaleht*, 8 November.

Tereblina, M. & Semyanov, A. (2002) *Narva arvudes 2002* [*Narva in figures 2002*] (Narva, Narva City Development and Economy Department).

Weiss-Wendt, A. (1997) *Must-valge linn. Vana Narva fotoajalugu* [*Black-and-white town. A photographic history of Old Narva*] (Tallinn, Tallinna Raamatutrükikoda).

ENVIRONMENT AND ENERGY: THE BALTIC SEA GAS PIPELINE

Ellen Karm

Germany and Russia have agreed to build the Nord Stream gas pipeline. The 1,200 kilometers-long structure will run from Vyborg, Russia to Greifswald, Germany traversing the Baltic Sea. The Baltic countries, Finland, Poland and Sweden, have expressed concerns about the project. At issue is the environmental impact of the project as well as the energy, political, and socioeconomic implications.

Germany and Russia have agreed to build a 1,200 kilometer underwater natural gas pipeline running from Vyborg to Greifswald through the Baltic Sea. Estonia, Latvia, Lithuania and Poland have spoken out against the project, expressing concern about its environmental and strategic ramifications. Other Baltic Sea region countries have also raised concerns. The pipeline will travel underwater through fishing and fish breeding grounds and through areas where chemical weapons from World War II were discarded. Furthermore, pollution and increased shipping have already affected the Baltic Sea, a fragile ecosystem.

The project – originally called the North European Gas Pipeline (NEGP), but now renamed Nord Stream – has the potential to harm the maritime environment, and consequently the Baltic Sea region countries. Other considerations include energy, inherent to the environmental concerns regarding the project's construction and operation, and its socioeconomic and political ramifications. International law is also an important factor. In this article I will argue that the Baltic Sea region countries and the EU should work together to safeguard the Baltic Sea environment and the surrounding countries from any anticipated environmental damage that could be caused by the construction and operation of the underwater pipeline. In particular, I argue the case for conducting an impartial impact study outlining environmental issues and the steps necessary to minimize risks. The Baltic Sea region countries already work together under the auspices of the Helsinki Convention, are members of the United Nations Economic Commission of Europe (UNECE), and have ratified the United Nations Convention on the Law of the Sea (UNCLOS).

In the case of the Nord Stream gas pipeline project, concerns have been raised regarding the environmental impact of the project, particularly the portion that will travel through the Baltic Sea. The countries that would be most impacted are located on the Baltic Sea and belong to the Helsinki Convention. Although no action was forthcoming for some time after the project was announced, the consortium, according to its website, has now filed its initial environmental notification complying with the UN's Espoo Convention. It anticipated completing its environmental impact assessment by mid-2007, and will seek final approval by 2008 (Nord Stream 2007). However, a number of Baltic Sea region countries and NGOs have continued to voice concerns over various aspects of the project. The UNECE Convention on Environmental Impact Assessment in a Transboundary Context, commonly known as the Espoo Convention, has been in force since 1997 and provides for notification and consultation among parties where a project like Nord Stream is likely to have transboundary environmental effects.[1] A relatively new treaty, it has been under study and there are conflicting opinions as to how effective it is in mitigating environmental harm, particularly in a multilateral context.[2]

It is my contention that if the affected parties are unable to address concerns regarding the project and reach an acceptable solution within the framework of existing agreements, a separate environmental agreement should be established to resolve problems connected with this project. Environmental treaties have grown considerably in number since the 1970s, and have become an essential part of international governance when it comes to resolving matters such as those raised by the Nord Stream gas pipeline project. As a number of recent studies in environmental policy theory have demonstrated, such treaties establish procedures, rules and social practices that govern and define the roles of the parties involved. Successful ones emphasize factors such as compliance, transparency, monitoring, sanctions and compensations (Young 1999; Miles 2002; Barrett 2003).

While the situation remains far from clear at present, a new environmental agreement may be necessary if those currently in existence do not provide the framework and mechanisms necessary to deal with the problems raised by Nord Stream – an agreement that utilizes, for instance, positive and negative incentives with respect to equipment standards, and the placement, construction and operation of the pipeline. Protection of wildlife and habitat, set-asides for fishing and fish breeding grounds will need to be considered, as well as compensation for any losses or damage that may result from the pipeline's construction and operation. The domestic law and policies of particular states will also come into play, regardless of whether the state in question stands to benefit from the natural gas that the pipeline will deliver or will be economically damaged by being bypassed while still subject to possible negative environmental effects. Who will provide leadership? Will coalitions form on one side or the other? If so, what would be their likely impact? These are factors that can be anticipated today, and the work toward resolving them can occur as the project moves forward.

Background

The Nord Stream pipeline project is a joint venture between the Russian state-owned Gazprom (51%) and two German companies, BASF AG (24.5%), a large chemical company, and E.ON AG (24.5%), an energy concern. Former German Chancellor Gerhard Schröder chairs the company shareholders committee.[3] This is a controversial position, since he accepted it shortly after leaving government, having strongly supported the project while chancellor. The pipeline project may be a relatively recent issue, but the underlying factors regarding its impact are not. The underwater portion to be built through the Baltic Sea has the potential to negatively affect its environment. Fish stocks, fisheries and discarded chemical munitions are important factors to consider in evaluating the environmental impact of the pipeline's construction, placement and operation. Besides the project's environmental impact this article will look at the issue of energy, the supply of which to Europe is the purpose of the Nord Stream project, and will also evaluate the political, economic and social considerations and their impact. Because of the limited scope of this article it is not possible to look at the impact on each individual country; thus, where appropriate the impact of the before-mentioned factors will be considered in relation to Estonia.

Ecological Characteristics of the Baltic Sea and Factors Affecting Marine Environment

One of the world's largest bodies of brackish water, the Baltic is a sheltered sea with natural access to the North Sea through the Danish Straits and Belt Sea. At an average depth of 56 meters, it is shallow with a slow rate of water exchange, taking over 20 years for a complete renewal of seawater. Its drainage area includes a population of 85 million, and it is one of the most polluted and stressed seas in the world, affected by eutrophication, agricultural wastes, contaminants from industrial activities and municipal waste discharges, increased shipping, overexploitation of living marine resources and invasive species (HELCOM 1998, pp. 1–3; MacKenzie *et al.* 2004).

Large basins separated by submarine elevations characterize its bottom and together with its narrow, natural access to the North Sea, ocean water high in saline and oxygen is restricted. The Baltic Sea has a permanent halocline or vertical gradient that prevents the mixing of the saltier water at the bottom and the fresher water in the upper levels. Therefore, water in the deeper areas of the sea has a higher saline content and less oxygen than in the surface or shallower areas. The coastal area is an important ecosystem, because of its naturally changing landscape and biodiversity. It also has high economic value with coastal cities and towns and industry. This can lead to conflict between those wishing to conserve nature and those wishing to exploit it (HELCOM 1998, pp. 1–4). In looking at any environmental impact from the construction and placement of the underwater pipeline, it is key to consider three factors that may be affected most by this project: chemical munitions dumped into the Baltic Sea after World War II, fish stocks and fisheries.

Discarded War Munitions

Chemical weapons, primarily German war materials, were dumped by American, British and Soviet militaries into the Baltic Sea after World War II. According to the Helsinki Commission (HELCOM), approximately 300,000 tons of chemical munitions were disposed of in this manner.[4] They included aerial bombs, mines, artillery shells and drums, and contained such toxic substances as adamsite, mustard gas, viscous mustard gas and tabun – a nerve gas. The amount of chemical agent in each weapon depended upon its type. For example, a bomb might contain 60% chemical agent, while an artillery shell might hold only 10% (Glasby 1997, pp. 267–71).

Around 26 to 40 German merchant ships loaded with munitions were sunk in the Baltic. Known dump areas are located east of Bornholm Island, south-east of Gotland Island and south of Little Belt. Not all the munitions were disposed of this way; some ships broke down prior to reaching their designated area, and boxes and crates containing weapons were reportedly tossed overboard en route. For example, wooden crates containing weapons were found washed ashore on the Swedish coast. Munitions are thus believed to be scattered throughout the Baltic Sea (Laurin 1991, p. 14; Glasby 1997, pp. 268–9).

Leakage of chemical agents has occurred as weapon casings have corroded over the years. Russian and Baltic scientists have found toxic substances in sediment, and gel-like clumps of mustard gas in the water and on the shore. Rusty, leaking shells along with mustard gas lumps have been caught in the nets of Danish and Swedish fishermen east of Bornholm Island (Laurin 1991, p. 14). Competition for fish may have compounded the problem. As fish stocks have been depleted, fishermen may have ventured into unsafe areas and employed more aggressive techniques such as bottom dragging. This can net a bomb or chemical agent along with the fish (Simons 2003). According to a HELCOM report prepared by Denmark, the lead country on this issue, there were four reported incidents of chemical weapons netted by fishermen in 2004, all near Bornholm Island. In 2003 there were 25 occurrences, and from 1995 to 2002 there were three to 11 incidents per year. Fishermen risk contamination from chemicals through touch or inhalation.[5]

There is uncertainty and disagreement between scientists, politicians, the military and NGOs regarding what should be done about the munitions dumps or scattered weapons, including over who should pay the cost of cleaning up or securing the area. Several solutions have been proposed. The Danish government concluded after numerous studies that the best solution would be to leave them alone. Some contend that bacteria will break down the chemicals over time. Covering the dumps with concrete or pumping it into the wrecks has also been proposed, but that solution would not work for the barrels, crates or shells that are reportedly scattered along the sea floor. Retrieval is a solution proposed by others, including former President of Estonia, Arnold Rüütel. Except for leaving the munitions alone, all other solutions are estimated to cost millions. HELCOM's 1996 report revealed no appreciable damage to the Baltic Sea, although they issued detailed instructions to fishermen on first aid equipment and treatment of contamination (Simons 2003). In 2005 HELCOM asserted that the chemical agents do not pose an extensive threat to the marine environment and the possibility of any threat to coastal areas is remote.[6]

The presence of discarded war materials and chemical munitions is nevertheless one of the factors that need to be studied and considered in the light of the pipeline project. Pipeline construction and operation should not exacerbate munitions pollution, and the munitions should not compromise pipeline safety. The German–Russian consortium has issued assurances that an environmental impact assessment will be conducted prior to beginning underwater construction. However, initial consultations have been criticized as inadequate (Deutsche Presse-Agentur 2007).

Fisheries and Distribution of Fish, Including Breeding Areas

The brackish water of the Baltic Sea means that it is habitable for only a few species, being too salty for some and too fresh for others. Different areas of the sea have different environmental and salinity conditions, making some areas more hospitable for certain species, but not others (Ojaveer & Lehtonen 2001). Cold-water species such as salmon, whitefish and trout were the first to migrate into the Baltic Sea. They continue to live in areas such as the Gulfs of Bothnia, Finland and Riga, where good oxygen conditions exist and they can reproduce at low temperatures. Marine species such as herring, cod and sprat found their way into the Baltic from the North Sea through the narrow passage by Denmark. The sea's low salinity has been the primary problem for northern marine species while low water temperature is a problem for southern species. Spring spawning herring, a marine species, is well adapted, reproduces in all areas of the Baltic, and maintains high populations. Cod and sprat are not as well adapted. Sprat cannot reproduce in areas with the lowest levels of saline such as the Gulf of Bothnia, and cod spawn only in the western and southern Baltic. Few species are able to reproduce in the north-eastern part of the sea (Ojaveer & Lehtonen 2001). Historically and through present day, freshwater fish are found in coastal areas of the Baltic and the rivers that flow into it. Indigenous species such as perch and pikeperch inhabit these areas along with cold-water fish adapted to low saline conditions. Freshwater species may vary from area to area, and are found in the West Estonian Archipelago and shallow areas of the coastal waters of the Gulfs of Bothnia, Finland, Riga and Gdansk. Obviously, fish stocks are dependent upon their ecosystem. In the second half of the twentieth century, human impact on fish stocks is evident. Fisheries in the Baltic primarily exploit cod, herring and sprat, which make up about 90% of the catch. In the 1990s fish stocks were variable. Sprat was plentiful, and good stocks of herring could be found in the large gulfs, but stocks were poor in the open sea areas of the southern and central Baltic. Cod may have experienced an increase, but this was countered by heavy exploitation. Salmon has also been depleted through overexploitation, damming of rivers and pollution. Mismanagement of fish stocks has resulted in overfishing like that of pikeperch in Pärnu Bay during the 1990s (Ojaveer & Lehtonen 2001). Overfishing can also reduce the numbers of large predatory fish with the result that smaller, inferior species such as gobies flourish.

Estonian Baltic Fishery

After independence, the eight cooperatives of the Estonian Baltic Sea fishery were privatized. The fleet consisted of a total 697 craft: 500 boats (12 meters or less) of the coastal fishery and 197 vessels of the offshore fishery. These craft were plagued by various problems including old age (75% were older than 10 years), poor maintenance and fuel economy, and outdated fishing equipment. The fish catch – primarily sprat and herring, but also cod – was about 60,000 tons annually for the years 1985 to 1991. However, in 1992 this dropped to 37,000 tons of which 30,000 tons were herring and 4,000 tons sprat (Food and Agriculture Organization 1994).

Table 1 shows the state of Estonia's Baltic Sea fishery for the years 2000 through 2005. The numbers are mixed; the Atlantic cod harvest in 2004 is more than twice what it was in 2000 and down by over half in 2005, while the trend for herring is down after showing slight improvement in 2004. The sprat catch is up considerably with a gain in 2005 of over 21,000 tons. The overall Baltic Sea catch shows improvement in 2004 and 2005 after a three-year decline. These figures appear hopeful, but it may be too early to tell if this is only a short-term reversal in a generally downward trend. According to HELCOM, cod and salmon in the Baltic Sea have declined as a result of overexploitation and pollution, and fish catches are not sustainable.[7] The construction and operation of the pipeline may place additional stress on fish populations and negatively impact the fishery by establishing off-limit areas to fishing.

Size and Path of Offshore Section of Pipeline and Supporting Structures

According to Gazprom and Nord Stream, the offshore section of the pipeline will run from Vyborg, Russia to Greifswald, Germany, with a coastal compressor station to be built in Portovaya Bay near Vyborg. The project will consist of two parallel pipelines,

TABLE 1 Estonian Baltic Sea fishery

Baltic Sea fish catch (measured in tons live weight)	Atlantic cod	Baltic herring	European sprat	Total Baltic Sea catch of all species measured
2000	514.2	41,735.0	41,393.8	85,176.0
2001	754.9	41,737.8	40,777.0	84,959.3
2002	37.2	36,250.3	40,717.1	79,034.8
2003	559.6	27,359.1	29,366.2	59,377.8
2004	1,278.7	27,380.0	34,113.4	64,902.3
2005	588.6	22,098.3	55,285.2	79,760.7

Source: Statistical Office of Estonia, updated 8 June 2006.

each of which will be 1,200 kilometers long and 1,219 millimeters in diameter. The annual capacity of the first pipeline will be 27.5 billion cubic meters (bcm), and the second pipeline will double that to about 55 billion cubic meters. The pipeline will be connected to the Russian gas transmission system. Possible spurs to areas further north and west in Europe have been proposed. Originally, it was to be completed by 2010, and expected to cost €4 billion (Gazprom 2005a). However, delays and cost overruns have reportedly moved up the expected completion date to 2011 and the cost to €6–8 billion. In 2007 Nord Stream announced a pipeline route change that would reroute the pipeline to the north of Bornholm Island, thus avoiding known weapon dump sites (Dempsey 2007b; World 2008). According to Nord Stream's website it is looking at route changes in three areas: the Gulf of Finland in Estonia's exclusive economic zone (EEZ), south of Gotland Island, and north of Bornholm Island (Nord Stream 2008).

According to scientist William Chandler, all pipelines have operational requirements. Transportation of natural gas through pipelines is done under pressure (tens of atmospheres). Requirements for long distance pipelines include compressor stations every 80–160 kilometers and gas release valves every 10–30 kilometers in the event of problems such as compressor failure (Chandler 2000, p. 109). A minimum number of supporting structures over the 1,200 kilometer route would require at least one compression station every 160 kilometers, resulting in seven or eight stations, plus a gas release valve every 30 kilometers, making a total of 40. If past experience is anything to go by, this could entail the loss of large areas to fishing. In 1986, for instance, British fishermen cited an independent report claiming that access to between 800 and 5,200 square kilometers of the North Sea had been lost as a result of offshore oil and gas activities. This represented an estimated £275,000–3,000,000 annually in lost revenue (Buchan & Allan 1992, p. 474).

Operation of Undersea Pipelines and Environmental Risk

Submarine pipelines of the offshore oil and gas industry represent a risk factor for environmental damage because the actual pipelines are susceptible to damage. They can be damaged by a ship's anchor and trawling nets, and are subject to defective materials and construction, corrosion and tectonic activity. Leakage from a damaged pipeline can be small or extensive depending upon how the harm was caused and its nature. Blowouts can occur, which may be explosive (Patin 1999, pp. 92–4).

Some countries have better safety and environmental records than others. During the period of oil and gas development in the North Sea there have been two serious accidents involving oil pipelines on the UK's continental shelf, as well as some minor spills and gas leaks from other North Sea pipelines. In 1980 a pipeline was damaged by a ship's anchor resulting in an oil spill, and in 1986 a spill occurred when a pipeline valve failed due to pressure changes (Haldane *et al.* 1992, pp. 519–20). In contrast, Russia's development of oil in Siberia and other locations has resulted in daily pipeline leakage (Patin 1999, p. 93). Standards and attitudes regarding the protection of the environment also play a role here. For example, despite the Soviet

Union's ratification of the Helsinki Convention in 1974, the Russian firm LUKOIL disregarded its requirements by drilling near Lithuania's Curonian Spit, a UNESCO World Heritage List site and an ecologically sensitive area. In 2004 appeals to Russia by the Baltic countries, Denmark, Finland and Sweden to stop the drilling were to no avail (Smith 2006, p. 43). Regardless of a country's track record, environmental damage can result if a spill or leak occurs. Nevertheless, a country's track record is a harbinger of its future behavior.

Law of the Sea

The United Nations Convention on the Law of the Sea of 10 December 1982 provides the legal framework under which global waters are defined as to ownership and use. Sections of the law pertinent to the pipeline concern territorial waters, continental shelves, exclusive economic zones, freedom of the high seas and protection and preservation of the environment. According to UNCLOS the territorial sea of a country extends 12 nautical miles (nm), normally measured from the low water line on its coast.[8] Freedom of the high seas applies to areas of the sea that are not part of an exclusive economic zone or in the territorial sea of a state. The high seas are open to all states and include the freedom to lay underwater pipelines and construct installations as allowable under international law.[9]

EEZs may be established up to 200 miles from the baseline of a country's coast and are measured in the same manner from the coast as the territorial sea measurement.[10] Furthermore,

> [A]ll States, whether coastal or land-locked, enjoy . . . the freedoms . . . of the laying of submarine cables and pipelines, and other internationally lawful uses of the sea related to these freedoms, such as those associated with the operation of ships, aircraft and submarine cables and pipelines, and compatible with the other provisions of this Convention.[11]

Paragraph 3 of Article 58 further specifies:

> In exercising their rights and performing their duties under this Convention in the exclusive economic zone, States shall have due regard to the rights and duties of the coastal State and shall comply with the laws and regulations adopted by the coastal State in accordance with the provisions of this Convention and other rules of international law in so far as they are not incompatible with this Part.[12]

Provision is also made for the resolution of conflicts between states in the EEZ 'on the basis of equity and in the light of all the relevant circumstances, taking into account the respective importance of the interests involved to the parties as well as to the international community as a whole'.[13]

Artificial installations and structures in the EEZ are covered by Article 60 of UNCLOS. According to its provisions the

> coastal State shall have the exclusive right to construct and to authorize and regulate the construction, operation and use of . . . (b) installations and structures for the purposes provided for in article 56 and other economic purposes;

(c) installations and structures which may interfere with the exercise of the rights of the coastal State in the zone.

The article further provides for notice of such construction and permanent warning, as well as reasonable safety zones around such installations and structures of not more than 500 meters for the safety of navigation and the structure. However, the zones should not be set where they would interfere with international sea lanes.[14]

UNCLOS also provides for cooperation of states bordering semi-enclosed seas, so as to protect and preserve the marine environment, and that cooperation would also extend to other states and international organizations.[15] Part XII provides for the protection and preservation of the marine environment, and establishes responsibilities and liabilities of states for damages.

The portions of the UN Law of the Sea that pertain to enclosed or semi-enclosed seas, territorial waters, exclusive economic zones, the high sea, laying of pipelines, as well as protection of the marine environment must be considered in light of the underwater pipeline project. Leaders in Estonia recently advocated extending the limits of the country's territorial sea to its legal limit. In 1993, Estonia and Finland voluntarily agreed to recede their respective territorial sea borders by three miles, creating a six-mile wide corridor as an EEZ that had the status of open sea regarding issues such as pipelines. The distinction is that a country has greater rights to protect its environment in a territorial sea as opposed to an EEZ (BBC Monitoring International Reports 2005). Estonia's territorial sea boundary is 12 nautical miles and the size of its EEZ is according to agreement with neighboring states.[16]

Impact of Natural Gas on Fish and Other Marine Organisms

According to scientist Stanislav Patin, there has been very limited study of the effect of natural gas and its components on water environments and organisms, and there are few scientific publications addressing the issue.

> Modern knowledge about the behavior of hydrocarbon gases in the natural waters and especially their impacts on water organisms, populations, and ecosystems is very limited. This, in turn, limits the possibility of prognoses, assessments, and prevention of environmental effects of accidental gas blowouts and leakages. (Patin 1999, p. 211)

However, after accidental drilling blowouts in the Sea of Azov in 1982 and 1985, researchers found a cause–effect relationship between mass fish mortality and the release of large amounts of natural gas. The fish showed 'impaired movement coordination, weakened muscle tone, pathologies of organs and tissues, damaged cell membranes, disturbed blood formation, modifications of protein synthesis, radically increased total peroxidase activity, and some other anomalies' that was indicative of fish poisoning (Patin 1999, pp. 235–6). Furthermore, according to an experimental study conducted with bottled gas, fish exhibited behavioral changes after exposure, and results seemed to indicate that methane and its derivatives had a greater effect on fish than zooplankton. Fish absorb gas primarily through their gills, and as with most species, the young are most vulnerable. Temperature and oxygen in the water environment are factors, because a rise in temperature increases the toxic effect while lower levels of oxygen in fish reduces their resistance to toxins. Exposure to gas may result in narcotic effects, damage the nervous system and tissues, and may cause death (Patin 1999, pp. 233–9). Natural gas can be released from a blowout or pipeline leak.

Natural gas is composed primarily of methane, up to 97%, and has a low rate of solubility, especially in seawater as compared to fresh water. The amount may vary, but coastal areas, bays and estuaries typically contain more methane naturally than the open ocean. In the water column, the distribution of methane occurs vertically, primarily toward the top of the column and sometimes at the bottom. It also occurs naturally from decomposing organic matter and seepage from formations containing oil or gas. (Methane, a greenhouse gas, is also implicated in climate change.) Other gases make up the remainder and may include hydrogen sulfide, which is water-soluble and can be highly toxic in both air and water (Patin 1999, pp. 220–2, 224–31, 249).

Assessment of Environmental Implications

All of the above-mentioned factors are salient when assessing the anticipated impact of the pipeline on the Baltic Sea environment. The building of the pipeline may disturb fish breeding grounds and disrupt fisheries in construction areas. Traditional fishing areas may be permanently displaced if a no fishing zone is implemented in the pipeline area to prevent just such damage. Gas leaks may harm or kill fish, their eggs, larvae and fry. Natural gas can also affect other marine organisms, such as zooplankton. The resulting damage would depend upon the location and extent of the gas leak as well as its duration and the proximity of fish stocks and breeding grounds.

Pipeline construction may also disturb chemical munitions, which are known to lie in the path of the pipeline. The possibility that natural gas might mix with chemical agents that have leaked from munitions also needs to be taken into account. In 2005 the consortium heading the project assured EU Energy Commissioner Andris Piebalgs that environmental studies would be done before construction commenced in the Baltic Sea (Agence France Presse – English 2005). However, it has been subsequently criticized for not seriously addressing environmental concerns. Nord Stream announced that during the first phase of international consultations it received 129 submissions from authorities and groups in Finland, Germany, Sweden, Denmark, the Baltic countries and Russia with respect to the environmental impacts of the pipeline (Deutsche Presse-Agentur 2007). The Baltic countries and Poland were the first to question the environmental impact of the project; more recently, these concerns have been reiterated by Sweden and Finland. In August 2006, for instance, Swedish Prime Minister Goran Persson voiced his concern that the pipeline would disturb discarded chemical weapons. Of particular interest to Sweden is the chemical weapons dump off Gotland Island (Osborn 2006). In February 2007 Finnish environmental officials also criticized the consortium's environmental assessments, calling them insufficient. They contended that the pipeline should be moved 15 kilometers to the south in the Gulf of Finland, where the seabed is more level and currents are more favorable to its construction and operation (BBC Monitoring European 2007b). The new path would place the pipeline outside of Finland's economic zone and into that of Estonia.[17] In this connection, Estonian Minister Juhan Parts met with officials from Nord Stream in May 2007 and expressed his concern about the pipeline's impact upon the environment and shipping lanes (Daily News Bulletin 2007). In September 2007 Estonia refused Nord Stream's request to survey the seabed located in the area encompassing its EEZ and territorial sea (BBC Monitoring Former Soviet Union 2007). The World Wildlife Fund (2004) has also raised environmental concerns about the project and its anticipated negative impact, especially on marine and bird species (Godoy 2006).

Assessment of Political, Socioeconomic and Energy Implications

The proposed Nord Stream pipeline has raised a whole host of complex national, regional and transnational environmental, political, economic, social and energy issues. The Baltic states for their part are worried about secure gas supplies. As they import gas from Russia, the pipeline will bypass them, as well as Poland, when delivering gas directly to western Europe. When Lithuanian President Valdus Adamkus met with German Chancellor Gerhard Schröder in October 2005, he criticized Schröder's indifference regarding the non-consultation of the Baltic countries over the project, as well as condemning the failure to address the issue at European Union level (Dempsey 2005). Latvian President Vaira Vīķe-Freiberga also claimed that Russia was willing to use energy policy to influence former Soviet republics politically, a concern also shared by Estonia and Lithuania (EIU ViewsWire 2006).

The Baltic Sea region countries have industries and lifestyles that could be impacted by environmental degradation related to pipeline construction or accidents with gas transport. One area that could be particularly affected is tourism, which has become a key growth industry in western Estonia (see Table 2). Total receipts expressed in US dollars (millions) are as follows: 367 in 2000; 504 in 2001;

TABLE 2 Accommodated tourists in Estonia

Year	2001	2002	2003	2004	2005
Entire country	1,227,000	1,402,000	1,562,000	1,922,000	2,073,000
Northern Estonia (Harju county including Tallinn)	783,000	808,000	865,000	1,124,000	1,200,000
Tallinn	759,000	775,000	826,000	1,063,000	1,131,000
Central Estonia (Järve, Lääne Viru, Rapla counties)	47,000	51,000	52,000	65,000	67,000
North-eastern Estonia (Ida Viru county)	12,000	58,000	78,000	71,000	74,000
Western Estonia (Hiiu, Lääne, Pärnu, Saare counties)	211,000	306,000	340,000	402,000	441,000
Southern Estonia (Jõgeva, Põlva, Tartu, Valga, Viljandi, Võru counties)	174,000	179,000	227,000	260,000	291,000

Source: Statistical Office of Estonia, 17 May 2006.

551 in 2002; 667 in 2003; and 806 in 2004.[18] Although the figures are not broken down according to specific areas of the country, it is reasonable to expect that income would be proportionate to numbers of visitors. According to Pärnu's tourism figures the number of visitors increased from a range of 249,000–275,000 in 2000 to a range of 275,000–304,000 in 2004. The number of visitors in the intervening years was even higher. Also the number of visitors to the Tourist Information Center increased annually from 37,000 in 2000 to 60,403 in 2004, and the number of hotel beds increased from 2,550 in 2000 to 3,548 in 2004 (Town of Pärnu 2005).

Energy is of increasing strategic importance today, both economically and politically, and is intrinsic to the German–Russian or Nord Stream underwater gas pipeline project. Also important is striking a balance between environmental concerns and the energy requirements of today's economies and lifestyles. With the growing economies of, for example, China and India and the accompanying increased global usage of oil and gas there is a corresponding increase in competition for the world's finite oil and gas resources. Countries with oil and gas reserves have historically used them to their economic and political advantage, and continue to do so. In his academic dissertation in 1997 Vladimir Putin spelled out his plan for Russia to regain its dominant global position by utilizing Russia's abundant natural resources, particularly its oil and gas reserves (Balzer 2006).

Gazprom, the Russian state-owned monopoly, has expanded its business throughout the European Union, taking advantage of Europe's liberalized energy markets by making deals with European energy companies and gaining access to their distribution pipelines and their customers. During the past several years Gazprom has acquired an increasing interest in the UK's energy industry, and has signed deals with Gaz de France, Eni of Italy and DONG Energy of Denmark. It also has distribution interests in Austria, Belgium and the Netherlands among others. In June 2006 Gazprom and Gasunie struck a deal that gave the Dutch firm an interest in the Nord Stream project while in turn Gazprom received an interest in Gasunie BBL or Balgzand-Bacton Line, a pipeline being built between the Netherlands and Britain. This gives Gazprom further entry into the UK market (Dempsey 2006; Platt's North Sea Letter 2006). Previously, Russia sought to increase its holdings including ownership of infrastructure and dominance of gas supplies through Gazprom's attempts to obtain control of the UK's largest gas company, Centrica, which is the parent company of British Gas. Although unsuccessful, Gazprom has reportedly not given up its goal of further entry into the UK's market. At the time, in response to Department of Trade and Industry actions to legally block the sale, Alexei Miller, the head of Gazprom, declared such efforts to check Gazprom's entry into the European market as politically motivated and not conducive to a positive outcome. He also warned about the competition for Gazprom's gas in other markets such as North America and China (BBC News 2006; Macalister 2007, p. 1). Gazprom's increasing interest in the European gas and energy market, especially the distribution part of the business outside its traditional role as a producer, causes uneasiness in those who worry that Russia through Gazprom is gaining too much control over this market which it may exploit for political gain.

The unreliability of supply and Russia's willingness to use energy as a political lever was exemplified by the disruption of oil and gas supplies during its disputes with Belarus and Ukraine. Early in 2007 oil deliveries were interrupted to Poland and Germany as a result of Russia's disagreement with Belarus, a transit country for its exports. As a result, Europeans again became uneasy about the reliability of Russian supplies, as they had been in January 2006 when natural gas was cut off to Europe during Russia's price dispute with Ukraine. While Gazprom is increasing its ties between production and distribution, European energy companies are eliminating theirs. At the same time Russia is denying European energy companies access to its gas pipelines (Dempsey 2007a).

The Sakhalin II project is another example of Russia's willingness to trample the rights of others in seeking gain for itself. Russia is accused of arbitrarily enforcing environmental regulations that it typically ignores in order to strong-arm Royal Dutch Shell, which owns a 55% stake in the project. After months of pressure, Shell sold a sizeable stake in the project to Gazprom. BP, the British company, ran afoul of Russian regulators over licensing technicalities because it had not developed the Kovykta gas field near China quickly enough. In June 2007 BP sold its controlling interest in the TNK-BP joint venture to Gazprom, once again the beneficiary of a pressured sale. Both Sahkalin II and TNK-BP are aimed at supplying the Asian market. The only other private energy firm in the region is Exxon Mobil, which is head of another joint venture and also under pressure from Russian environmental officials. According to energy analysts, Russia's aim is to dominate the market, and it is not averse to playing the European and Asian markets against each other for its own gain (Knight Ridder Tribune Business News 2007; Kramer 2007a, 2007b).

The EU's response to these developments has been low-key, which may have to do with the fact that Russia supplies a quarter of its natural gas (Commission of the European Communities 2006b, Annex, p. 24). However, Russia's actions have been criticized in the press and it has been reminded of its need for foreign investment and expertise, which may be curtailed in the future if foreign investors find that their rights are not protected, but rather subject to political expediency (*The New York Times* 2006). Rather than diversifying Europe's gas supply, a goal of liberalization, Gazprom is gaining a greater presence, which may result in decreased competition and market access.

Russia's efforts to gain control of energy companies in the UK and gas projects aimed at the Asian market mirror its actions during the 1990s in the Baltic states. Russia sought political influence in Lithuania through the Russian oil company LUKOIL, which was involved in discussions with the Lithuanian government and the American firm Williams International Company for an interest in the Lithuanian oil complex Mazeikiai Nafta in the late 1990s. An agreement was eventually reached with Williams and the then privately owned YUKOS; however, during protracted negotiations Russia cut off oil supplies to Lithuania nine times from 1998–2000 (Clemens 2001, pp. 145–7; Smith 2006, pp. 41–2). Russia has further sought political influence through controlling or major stakes owned by Russian energy companies in all three Baltic countries. For instance, Gazprom owns the majority interest of 37% in the Estonian gas trading and transport company Eesti Gaas; a 25% stake in Latvijas Gāze, a Latvian gas trading and transport business; and 30% in the Lithuanian gas trading firm Stella Vitae (Smith 2006, p. 32).

Thus, in terms of energy politics, Russia continues to be a dominating force as far as the Baltic states are concerned. Estonia and the others would benefit from a cooperative, friendly relationship with Russia, but not at the expense of political domination. Because of its small size, Estonia's best path toward secure energy supplies is through its European Union membership. This was the path advocated by Estonia's President Toomas Hendrik Ilves – then deputy chairman of the European Parliament foreign affairs committee – who characterized the Russian–Ukrainian gas dispute as political in a January 2006 speech. According to Ilves, all of Europe faced increased risk without a definite energy security policy (Baltic News Service 2006a). Later, he underlined the importance of a common energy policy in securing the EU's energy supply, arguing that it could mirror the success achieved by the EU's common trade policy by negotiating from a stronger united position versus bilateral agreements negotiated by each member country (Baltic News Service 2006b).

A step forward in EU solidarity occurred in November 2006, when negotiations between the EU and Russia for a new treaty of cooperation including energy, justice, and security among other concerns was blocked by Poland's veto over Russia's ban on its agricultural products. In a January 2007 interview European Commission President Jose Manuel Barroso characterized the stalled negotiations as exhibiting the EU's strength and solidarity, and that solutions would be sought to safeguard Poland's interests while moving forward with talks with Russia. With respect to the concerns of Poland and the Baltic countries that Russia would use its energy resources for political intimidation, Barroso emphasized the importance of the EU having a common energy policy, which would be beneficial economically and in terms of security. However, he also stressed the importance of the energy issue for EU–Russia relations and the interdependency of the two parties in this regard (BBC Monitoring European 2007a). In April Poland removed its objection and the European Commission announced that negotiations would proceed with Russia with progress dependent upon both parties (BBC Monitoring European 2007c).

Table 3 includes natural gas statistics for the Baltic countries, and other near neighbors who are also EU members including Germany and Russia. The table helps to illustrate the importance of natural gas supplies to all of these countries. As is apparent, Russia has vast gas reserves. Of the Baltic Sea region countries that are EU members only Denmark is wholly natural gas independent and exports more than it uses domestically. Germany and Poland also have gas reserves, but they are insufficient for their needs. All the other countries are completely dependent on outside sources for their natural gas supplies. This situation can make them vulnerable to outside political and economic influence.

As shown in Table 3, Estonia must import natural gas to meet its needs. In 2001 it imported 29,803 terajoules; 24,963 in 2002; and 27,509 in 2003 (Statistical Office of Estonia 2006). Electricity is generated primarily by oil shale and to a small extent by hydroelectric power and more recently by wind power. Oil shale is a source that is domestically available and independent of outside influence, but is also

TABLE 3 Natural gas statistics

Country	Natural gas production	Natural gas consumption	Natural gas exports	Natural gas imports	Natural gas proved reserves
Denmark	9.87 bcm (2006 est.)	4.824 bcm (2006 est.)	4.964 bcm (2006 est.)	0 cm (2006 est.)	99.99 bcm (2006 est.)
Estonia	0 cm (2004 est.)	1.44 bcm (2004 est.)	0 cm (2004 est.)	1.44 bcm (2004 est.)	–
Finland	0 cm (2004 est.)	4.86 bcm (2004 est.)	0 cm (2004 est.)	4.866 bcm (2004 est.)	–
Germany	19.9 bcm (2005 est.)	102 bcm (2005 est.)	8.81 bcm (2004)	90.11 bcm (2004 est.)	279.1 bcm (1/1/05 est.)
Latvia	0 cm (2004 est.)	1.8 bcm (2003 est.)	0 cm (2001 est.)	1.7 bcm (2001 est.)	–
Lithuania	0 cm (2004)	2.92 bcm (2004 est.)	0 cm (2004 est.)	2.92 bcm (2004 est.)	–
Poland	5.957 bcm (2004)	15.67 bcm (2004 est.)	46 mcm (2004 est.)	9.963 bcm (2004 est.)	164.8 bcm (01/01/04 est.)
Russia	641 bcm (2005 est.)	445.1 bcm (2005 est.)	216.8 bcm (2004 est.)	36.6 bcm (2004 est.)	47.57 tcm (01/01/03)
Sweden	0 cm (2004 est.)	979 mcm (2004 est.)	0 cm (2004 est.)	979 mcm (2004 est.)	–

Source: US Central Intelligence Agency, *The World Factbook*, 15 May 2007.

Note: Cubic meters (cm), million cubic meters (mcm), billion cubic meters (bcm), trillion cubic meters (tcm).

environmentally polluting. Consumption of oil shale has been rising as well from (expressed in thousands of tons) 13,077 in 2002; 15,474 in 2003; and 15,501 in 2004. During that time, production of hydro and wind power also increased from 26 terajoules in 2002 to 68 in 2003 and 108 in 2004 (Clemens 2001; Statistical Office of Estonia 2006). Although such power sources represent a fraction of the whole, they are a step in the right direction toward environmental and energy sustainability. Increased sustainable energy production would benefit Estonia environmentally, politically and economically. For example, oil shipments were abruptly halted by the state-owned Russian Railways in spring 2007 during the 'bronze soldier' dispute between Estonia and Russia (Halpin 2007). As a relatively new and an economically, geographically and demographically small EU member, its influence on EU policy reflects these realities.

Although the EU has long had a common environmental policy, member countries have not been willing to give up their sovereignty in the area of energy. However, this may be changing. Europeans were greatly troubled by the abrupt cut-off of energy supplies during Russia's dispute with Ukraine and Belarus. Also of concern has been Russia's move away from transparent democratic policies. For example, the repression of independent media, the restrictive law governing NGOs, as well as the crackdown on pro-democracy demonstrations. Moreover, the recent law signed by Putin would put additional pressure on political opposition (Schwirtz 2007). As discussed above, Gazprom's stepped-up actions to increase its energy holdings in Europe have also caused concern. The time may be right for a policy change.

In March 2006 the Commission of the European Communities issued a Green Paper entitled *A European Strategy for Sustainable, Competitive and Secure Energy.* The paper advocates a common course citing the insufficiency of individual energy policies. It further asserts that the EU, as the world's second largest energy market, has the clout to protect its interests if member states act together. It urges innovation and diversity as to the type of energy, its transit and country of origin. The paper identifies six key areas where action is needed and where such a common strategy would encompass the core principles of sustainability, competitiveness and security. Those areas are competitiveness and the internal market, diversification, solidarity, sustainable development, innovation and technology, and external policy (Commission of the European Communities 2006a, pp. 3–5). Europeans seem to support such a policy change. According to a 5 March 2007 Eurobarometer survey, 62% of EU citizens prefer common EU decisions regarding energy and 65% believe the EU is better able to negotiate energy supplies and prices versus member states doing so on their own.[19]

In the *Annex to the Green Paper*, the section on natural gas reaches several conclusions with respect to EU energy policy. First, consumption of gas has grown over the years, and its use can provide environmental and economic advantages in comparison to other forms of energy. Second, EU production has to date accounted for 46% of its gas supplies, Russia 25%, and Norway along with others making up the remainder. However, this situation will change in the future due to declining UK production and limited reserves, a 50% reduction over the next 20 years, and an annual increase in imports from 275 bcm to 535 bcm by 2030 is anticipated. Third, the EU views diversification of supply to be of utmost importance to security of supply. For import supplies over the next five to ten years it is depending on a number of projects that are planned or in the beginning phases of construction, for example, the gas pipeline through the Baltic Sea. Finally, the Green Paper sees gas and its price as major factors in the EU's competitiveness as well as having a positive environmental impact in terms of greenhouse gas emissions if used as a replacement for other more polluting fuels (Commission of European Communities 2006b, Annex, pp. 24–7). In 2007 the EU proposed a new energy policy that reiterates points set forth in its 2006 Green Paper. It further emphasizes solidarity between member countries and security of supply plus diverse sources of supply and of transport routes as well as separate ownership of network management, production and sales in the internal market.[20]

As set forth in *Annex to the Green Paper*, the proposed EU common energy policy apparently takes it as given that the underwater gas pipeline in the Baltic Sea will be constructed. Politically it would be difficult for Estonia, Latvia, Lithuania or Poland to halt this project. According to the newspaper, *Postimees*, Polish Defense Minister Radek Sikorski, during a meeting with German Chancellor Angela Merkel, asked Germany to give up the Baltic Sea underwater pipeline project with Russia, but Germany refused. Merkel offered to discuss the matter further with Poland, but the Polish position was that the matter should have been discussed before the pipeline project agreement with Russia was finalized (*Postimees* 2006). In light of this, and from an environmental standpoint, the best course of action is to take steps to ensure the project causes the minimum amount of disruption to the Baltic Sea environment and the surrounding region as possible.

Conclusion

The Baltic Sea region countries should work together to safeguard the Baltic Sea environment and the surrounding countries from any anticipated environmental damage that could be caused by the construction and operation of the underwater pipeline. An impartial impact study should be conducted taking into account the factors outlined in the article, and steps taken to minimize risks. Such oversight could help assure the integrity of the pipeline while minimizing the environmental impacts of its construction and operation in consideration of the Baltic Sea's unique and fragile ecosystem. As proposed in the introduction to this article, if the parties impacted by this project are unable to address concerns regarding it and reach an acceptable solution within the framework of existing agreements, a separate environmental agreement should be established to investigate and resolve the problems connected with this project. Environmental treaties have become an essential part of international governance in resolving matters such as those raised by the Nord Stream gas pipeline.

All Baltic Sea region countries have a stake in ensuring that this project will not cause harm to the environment. The issue of the pipeline is not just a bilateral matter between Russia and Germany, or the other countries that have struck bilateral deals with Gazprom, but affects all Baltic Sea region countries. Despite Russian and German assurances of safety and a comprehensive environmental impact study prior to the start of offshore construction, initial work has been called inadequate. The Baltic countries, Finland, Poland and Sweden have come out against or expressed their concerns about the environmental soundness of the underwater pipeline project. Environmental degradation resulting from the pipeline could negatively impact Baltic Sea fisheries, fish stocks, disturb chemical weapons discarded decades ago, and further compromise an already stressed ecosystem.

This project also has political and economic ramifications for all those affected by it. Since Putin's arrival in power, Russia has sought to regain influence in world affairs through the exploitation of its oil and gas reserves. Germany with the largest economy and second largest population in Europe is largely dependent upon outside sources to meet its energy needs.[21] Natural gas is a relatively clean fuel and can help meet EU common environmental standards. A bilateral agreement allows Germany to negotiate for its own energy requirements without having to consider how that agreement would affect other EU members. The same is true for the other European countries where bilateral deals have been made with Gazprom for gas from the Nord Stream pipeline. The Baltic states and Poland rely wholly or to a large extent upon Russia for their gas supplies. The Baltic countries have also served as transit countries for Russian energy supplies headed for western Europe. Their position as part of the delivery process has afforded them income and a measure of security (BBC Monitoring Former Soviet Union – Economic 2002; Smith 2006, pp. 31–2, 43–6). Russia, however, now argues that transit countries are barriers to reliable gas deliveries to western Europe and bypassing them via the Nord Stream pipeline will solve that problem. However, it is Russia that has commonly cut off supplies as a form of political pressure. By painting the Baltic states and other former Soviet bloc countries as barriers to the free flow of gas to western Europe it is attempting to drive a wedge between Western and Eastern

EU members. Indeed, it has argued that these former Soviet bloc countries are responsible for the recent souring of its relations with the EU (Financial Times 2007).

In light of the factors set forth in this article it is apparent that the environmental effects of this project cannot be evaluated alone. Energy and the environment are two sides of the same coin and they must be considered by looking at their ramifications on domestic and multinational socioeconomic and political factors and their strategic implications. Germany appears to view Russia as an indispensable partner for its gas supplies, as does the EU. However, Russia lacks the environmental laws and citizen involvement, that have been crucial to enforcement of Western environmental standards. It also lacks political and economic transparency (Smith 2006, pp. 75–6). It has repeatedly demonstrated its willingness to gain political influence in Europe through its oil and gas resources and the acquisition of supporting infrastructure and industries.

In 2006 the European Union published a Green Paper proposing a common energy policy to replace the 27 separate energy policies that now exist, and legislation for such a policy was proposed in 2007. Polls show that a large percentage of Europeans favor a common energy policy. The Baltic states would be best served by a common EU energy policy, which could afford them more energy and political security. The actions of Germany, Italy, the Netherlands and other EU member countries in striking bilateral agreements with Russia for gas will satisfy their short-term economic and political needs. However, in the long term these deals will diminish the economic and strategic power of the EU. As a united community, the EU has substantial political and economic clout, which should grow as the former Soviet bloc countries, the EU's newest members, grow and refine their democracies and economies to catch up with the western European nations. This power will lessen if the EU allows itself to be fragmented by energy issues that in turn impact environmental, economic, social, and political ones. Democracy in Russia has been eroded under Putin's leadership, which has been characterized by saber rattling against internal and external opponents, and increased assertiveness in relation to the former Soviet republics. The EU should not allow itself to be divided by political and economic shortsightedness and Russia's determination to exert its power on Europe's stage through its energy resources. It should adopt a common energy policy that would benefit all its members. As stated before, environment and energy go hand-in-hand and are interconnected to politics and socioeconomics. For too long these issues have been dealt with piecemeal. A comprehensive solution would depend upon a common approach that would mitigate or solve the environment–energy problem in the long term for all those concerned.

Acknowledgements

This article grew out of a research paper and a chapter of my master's thesis. My profound thanks to Christine Ingebritsen, Vladimir Kaczynski, Guntis Smidchens and Nathaniel Trumbull. Thanks also to the anonymous reviewers for their comments and suggestions.

Notes

1 The convention, in force since 1997, is available at http://www.unece.org/env/eia/documents/conventiontextenglish.pdf, accessed 14 December 2007.
2 For differing views of the effectiveness of the Espoo Convention see Koivurova (2007); Knox (2002); and Albergaria and Fidelis (2006).
3 Gazprom (2005b); see also Anna Shiryaevskaya and Nadia Rodova (2005).
4 HELCOM, 'Helcom Executive Secretary Anne Christine Brusendorff Interview with Deutsche Welle Radio on WWII Chemical Munitions Dumped in the Baltic Sea', *News*, available at: http://www.helcom.fi/press_office/news_helcom/en_GB/Interview.11156, accessed 13 March 2006.
5 HELCOM, 'Statistics Shows Significant Drop in the Number of Incidents of Chemical Munitions Caught by Fishermen in the Baltic Sea', *News*, 2 May 2005, available at: http://www.helcom.fi/press_office/news_helcom/en_GB/ChemicalMunitions1115039886140, accessed 15 May 2007.
6 *Ibid.*
7 HELCOM, *Fisheries*, available at: http://www.helcom.fi/environment/assessment/fish/en_GB/fisheries, accessed 28 May 2007.
8 United Nations, *United Nations Convention on the Law of the Sea of 10 December 1982* – overview and full text (UNCLOS), last updated 16 March 2006, available at: http://www.un.org/Depts/los/convention_agreements/texts/unclos_e.pdf, Part II, Section 2, Articles 3–5, accessed 26 April 2006.
9 *Ibid.*, Part VII, Section 1, Article 87.
10 *Ibid.*, Part V, Article 57; see also HELCOM (2000, p. 1).
11 UNCLOS, Part V, Article 58.
12 *Ibid.*
13 *Ibid.*, Article 59.
14 *Ibid.*, Article 60.
15 *Ibid.*, Part IX, Article 123 (b) and (d).
16 US Central Intelligence Agency, *The World Factbook*, available at: http://www.cia.gov/cia/publications/factbook/geos/en.html, accessed 28 May 2007.
17 Office of the President of the Republic of Estonia, 'Presidents Halonen and Ilves: A Strong European Union is in Finland's and Estonia's Interests', *Press Releases*, 14 March 2007, available at: http://www.president.ee/en/duties/press_sreleases.php?gid=90894, accessed 15 May 2007.
18 World Tourism Organization, *Tourism Market Trends*, 2005 *Edition – Annex* (data collected November 2005), available at: http://www.world-tourism.org/facts/menu.html, accessed 3 June 2006.
19 European Union, 'Eurobarometer: Europeans Support Greater EU Action on Energy and Climate Change', *EUROPA-Rapid-Press Releases*, 5 March, available at: http://europa.eu/rapid/pressReleasesAction.do?reference=IP/07/280&format=html&aged=1&language=EN&guiLanguage=en, accessed 28 February 2008.
20 European Union, 'An Energy Policy for Europe', 20 November 2007, available at: http://europa.eu/scadplus/leg/en/lvb/127067.htm, accessed 29 January 2008.
21 US Central Intelligence Agency, *The World Factbook*, 15 May 2007, available at: http://www.cia.gov/library/publications/the-world-factbook/geos/gm.html, accessed 28 May 2006.

References

Agence France Presse – English (2005) 'Baltic Sea Pipeline Won't be Built Without Environment Study: EU', 5 December, available at: http://web.lexis-nexis.com, accessed 26 January 2006.

Albergaria, R. & Fidelis, T. (2006) 'Transboundary EIA: Iberian Experiences', *Environmental Impact Assessment Review*, 26, pp. 614–32.

Baltic News Service (2006a) 'Estonian ME: EU Should Deal with Energy Policy Risks', 17 January, available at: http://web.lexis-nexis.com, accessed 15 April 2006.

Baltic News Service (2006b) 'Estonia MEP Underlines I[m]portance of Common EU Energy Policy', 23 March, available at: http://web.lexis-nexis.com, accessed 15 April 2006.

Balzer, H. (2006) 'Vladimir Putin's Academic Writing and Russian Natural Resource Policy', *Problems of Post-Communism*, 53, 1, January/February, pp. 48–54.

Barrett, S. (2003) *Environment and Statecraft: The Strategy of Environmental Treaty-Making* (Oxford, Oxford University Press).

BBC Monitoring European (2007a) 'EU's Barroso Tells Lithuanian Weekly: Connecting Energy Markets "Priority"', 5 January, available at: http://proquest.umi.com, accessed 27 April 2007.

BBC Monitoring European (2007b) 'Finnish Daily on Russian–German Baltic Sea Natural Gas Pipeline Project', 5 March, available at: http://proquest.umi.com, accessed 6 March 2007.

BBC Monitoring European (2007c) 'Poland Premier States Conditions for Lifting Veto on EU–Russia Talks', 4 April, available at: http://proquest.umi.com, accessed 27 April 2007.

BBC Monitoring Former Soviet Union – Economic (2002) 'Estonian Newspaper Warns Russia's Growing Oil Trade Will Endanger Baltic Ports', 10 January, available at: http://proquest.umi.com, accessed 27 November 2005.

BBC Monitoring Former Soviet Union (2007) 'Estonia's Envoy to Russia Explains why Survey Request was Rejected', 20 September, available at: http://proquest.umi.com, accessed 22 January 2008.

BBC Monitoring International Reports (2005) 'Four Estonian Public Figures Advocate Extension of Territorial Sea Border', 30 December, available at: http://web.lexis-nexis.com, accessed 19 January 2006.

BBC News (2006) 'Gazprom Warns EU to Let it Grow', 20 April, available at: http://news/bbc.co.uk/go/pr/fr/-1/hi/business/4925682.stm, accessed 26 April 2006.

Buchan, G. & Allan, R. (1992) 'The Impact on Fishing Industry', in Cairns, W. (ed.) (1992) *North Sea Oil and the Environment: Developing Oil and Gas Resources, Environmental Impacts and Responses* (London, Elsevier Applied Science), pp. 459–80.

Chandler, W. (2000) *Energy and Environment in the Transition Economies* (Boulder, Westview Press).

Clemens Jr, W. C. (2001) *The Baltic Transformed: Complexity Theory and European Security* (Lanham, Rowman & Littlefield).

Commission of the European Communities (2006a) *Green Paper: A European Strategy for Sustainable, Competitive and Secure Energy*, SEC (2006), 317, Brussels, 8 March, available at: http://europa.eu.int/comm/energy/green-paper-energy/doc/2006_03_08_gp_document_en.pdf, accessed 15 April 2006.

Commission of the European Communities (2006b) Commission Staff Working Document – Annex to the *Green Paper: A European Strategy for Sustainable, Competitive and Secure Energy – What is at Stake – Background Document*, SEC (2006) 317/2, Brussels, 8 March, available at: http://europa.eu.int/comm/energy/green-paper-energy/doc/2006_03_08_gp_document_en.pdf, accessed 15 April 2006.

Daily News Bulletin (2007) 'Estonian Minister Urges Nord Stream to Consider Alternative Routes', 4 May, available at: http://proquest.umi.com, accessed 12 July 2007.

Dempsey, J. (2005) 'Lithuanian Leader Faults EU over New Gas Pipeline Common Strategy is Lacking, He Asserts', *International Herald Tribune*, 27 October, available at: http://proquest.umi.com, accessed 4 December 2005.

Dempsey, J. (2006) 'Gazprom Strikes Deal with Dutch; [3 Edition]', *International Herald Tribune*, 9 June, p. 18, available at: http://proquest.umi.com, accessed 9 February 2007.

Dempsey, J. (2007a) 'Gazprom Exploits EU Disarray Energy Policy is Not Unified among Very Different Nations; [1R Edition]', *International Herald Tribune*, 12 January, p. 9, available at: http://proquest.umi.com, accessed 9 February 2007.

Dempsey, J. (2007b) 'Gazprom Plans Shift of European Pipeline Route Mitigates Environmental Issues; [5 Edition]', *International Herald Tribune*, 24 August, p. 3, available at: http://proquest.umi.com, accessed 22 January 2008.

Deutsche Presse-Agentur (2007) 'NordStream to Conduct More Studies for Baltic Gas Pipeline', 4 April, available at: http://rigzone.com/news/article_pf.asp?a_id=43496, accessed 19 April 2007.

EIU ViewsWire (2006) 'Estonia/Latvia/Lithuania Politics: In NATO, but Still Nervous', 30 November, available at: http://proquest.umi.com, accessed 20 May 2007.

European Union (2007) 'Eurobarometer: Europeans Support Greater EU Action on Energy and Climate Change', *EUROPA-Rapid-Press Releases*, 5 March, available at: http://europa.eu/rapid/pressReleasesAction.do?reference=IP/07/280&format=HT . . ., accessed 26 April 2007.

Financial Times (2007) 'Gazprom to Press on with Investment in EU', 23 May, available at: http://www.ft.com/cms/s/f41cfe12-08ca-11dc-b11e-000b5df10621,dwp_uuid=4e612cca-6706-. . ., accessed 3 June 2007.

Food and Agriculture Organization of the United Nations (1994) 'Review of the Estonian Fishery Sector (Report and Annexes)', available at: http://www.fao.org/docrep/field/373971.htm, accessed 1 May 2006.

Gazprom (2005a) 'Importance: The North European Gas Pipeline (NEGP)', available at: http:www.info/news/news1.htm, accessed 22 February 2006.

Gazprom (2005b) 'The North European Gas Pipeline (NEGP), 2005', available at: http://www.info/news/news16.htm, accessed 22 February 2006.

Glasby, G. P. (1997) 'Disposal of Chemical Weapons in the Baltic Sea', *The Science of the Total Environment*, 206, August, pp. 267–73.

Godoy, J. (2006) 'Europe: Gas Pipeline Threatens Fragile Marine Ecosystems', *Global Information Network*, 22 June, available at: http://proquest.umi.com, accessed 18 July 2006.

Haldane, D., Paul, M. A., Reuben, R. L. & Side, J. C. (1992) 'Submarine Pipelines and the North Sea', in Cairns, W. (ed.) (1992) *North Sea Oil and the Environment: Developing Oil and Gas Resources, Environmental Impacts and Responses* (London, Elsevier Applied Science), pp. 481–522.

Halpin, T. (2007) 'Russia Cuts Off Oil in Battle over War Statue', *The Times*, 3 May, available at: http://proquest.umi.com, accessed 5 June 2007.

HELCOM (1998) *Red List of Marine and Coastal Biotopes and Biotope Complexes of the Baltic Sea, Belt Sea and Kattegat – Including a Comprehensive Description and Classification System for All Baltic Marine and Coastal Biotopes – Balt. Sea Environ. Proc. No.* 75 (Helsinki, Helsinki Commission).

HELCOM (2000) *Baltic Legal Manual – Information on Anti-pollution Regulations at Sea and the Prosecution of Violations Thereof in the Baltic Sea Area, Baltic Sea Environ. Proc. No.* 77 (Helsinki, Helsinki Commission).

Knight Ridder Tribune Business News (2007) 'Gazprom Gets Majority Stake in Sakhalin-2 Project', 19 April, p. 1, available at: http://proquest.umi.com, accessed 26 January 2008.

Knox, J. (2002) 'The Myth and Reality of Transboundary Environmental Impact Assessment', *The American Journal of International Law*, 96, 2, April.

Koivurova, T. (2007) 'The Convention on Environmental Impact Assessment in a Transboundary Context (Espoo Convention)', in Ulfstein, G., Marauhn, T. & Zimmermann, A. (eds) (2007) *Making Treaties Work* (Cambridge University Press, Cambridge).

Kramer, A. E. (2007a) 'Kremlin Aim of Monopoly Threatens BP Gas Venture', *The New York Times*, 29 May, p. C3.

Kramer, A. E. (2007b) 'The Kremlin Flexes, and a Tycoon Reels', *The New York Times*, 8 July, Sec. 3, pp. 1, 6.

Laurin, F. (1991) 'Scandinavia's Underwater Time Bomb', *Bulletin of the Atomic Scientists*, 47, 2, pp. 10–15.

Macalister, T. (2007) 'Gazprom Targets Key UK Names in Drive for Expansion', *The Guardian*, 12 April, available at: http://proquest.umi.com, accessed 27 April 2007.

MacKenzie, B. R., Almesjo, L. & Hansson, S. (2004) 'Fish, Fishing, and Pollutant Reduction in the Baltic Sea', *Environmental Science and Technology*, 38, 7, February, pp. 1970–6, available at: http://pubs.acs.org/cgi-bin/article.cgi/esthag/2004/38/i07/html

Miles, E. L. (2002) 'Sea Dumping of Low-level Radioactive Waste', in Choucri, N. (ed.) (2002) *Environmental Regime Effectiveness: Confronting Theory with Evidence* (Cambridge, MIT Press), pp. 87–116.

The New York Times (2006) 'The Kremlin's Shell Game', 19 December, p. A30.

Nord Stream (2007) 'Environmental Impact Assessment', available at: http://www.nord-stream.com/eng/ecology/impact_assessment/, accessed 23 May 2007.

Nord Stream (2008) 'Route', available at: http://www.nord-stream.com/route.html, accessed 30 January 2008.

Ojaveer, E. & Lehtonen, H. (2001) 'Fish Stocks in the Baltic Sea Finite or Infinite Resource?' *Ambio: A Journal of the Human Environment*, 30, 4, pp. 217–21, available at: http://www.bioone.org/bioone/, accessed 14 February 2006.

Osborn, A. (2006) 'Baltic Pipeline will Lead to Disaster, Sweden Warns', *Belfast Telegraph*, 24 August, p. 1.

Patin, S. (1999) *Environmental Impact of the Offshore Oil and Gas Industry* (translated by E. Cascio) (East Northport, EcoMonitor Publishing).

Platt's North Sea Letter (2006) 'Gazprom Signs Long-Term Gas Supply Deal with DONG', 21 June, available at: http://proquest.umi.com, accessed 5 June 2007.

Postimees (2006) 'Poola: Merkel keeldub Vene gaasijuhtmest loobumast', 30 April, available at: http://www.postimees.ee/020506/esileht/valisuudised/199896_print.php, accessed 9 May 2006.

Schwirtz, M. (2007) 'Russia: Putin Signs Amendments on Extremism', *The New York Times*, 27 July A6.

Shiryaevskaya, A. & Rodova, N. (2005) 'Gazprom Begins Building Direct Gas Pipeline Route to Europe via Baltic Sea', *Platts Oilgram News*, 12 December, available at: http://web.lexis-nexis.com, accessed 8 February 2006.

Simons, M. (2003) 'Discarded War Munitions Leach Poisons into the Baltic', *The New York Times*, 20 June, sec. A, available at: http://proquest.umi.com, accessed 27 November 2005.

Smith, K. C. (2006) *Russian Energy Politics in the Baltics, Poland, and Ukraine: A New Stealth Imperialism?* (Washington, DC, Center for Strategic & International Studies).

Statistical Office of Estonia (2006) *The Baltic Sea Catch by Economic activity, Year and Species*, last updated 8 June 2006, available at: http://pub.stat.ee/px-web.2001/Dialog/Saveshow.asp, accessed 26 April 2007.

Town of Pärnu (2005) *Pärnu in Figures 2004*, 27 September, available at: http://www.Parnu.ee/?id=416, accessed 25 April 2006.

World, C. M. (2008) 'Severe Delays and Rising Costs Hamper Baltic Sea Gas Pipeline; [Final 1 Edition]', *The Times*, 7 January, available at: http://proquest.umi.com, accessed 22 January 2008, p. 37.

World Wildlife Fund (2004) 'WWF Baltic Ecoregion Conservation Plan', *Publications*, 25 May, Executive Summary, available at: http://www.panda.org/about_wwf/where_we_work/europe/what_we_do/baltics/publication . . ., accessed 12 April 2007.

Young, O. (1999) 'Hitting the Mark: Why Are Some International Environmental Agreements More Successful than Others?', *Environment*, 41, 8, pp. 20–9.

Index